Reining in the Rio Grande

Reining in the Rio Grande

People, Land, and Water

Fred M. Phillips

G. Emlen Hall

Mary E. Black

University of New Mexico Press | Albuquerque

Library of Congress Cataloging-in-Publication Data
Phillips, Fred M. (Fred Melville)
  Reining in the Rio Grande : people, land, and water / Fred M. Phillips, G.
Emlen Hall, Mary E. Black.
     p. cm.
  Includes bibliographical references.
  ISBN 978-0-8263-4943-9 (cloth : alk. paper)
  1. Human ecology—Rio Grande Valley. 2. Riparian areas—Rio Grande
Valley. 3. Riparian ecology—Rio Grande Valley. 4. Riparian restoration—
Rio Grande Valley. 5. Irrigation farming—Rio Grande Valley. 6. Rio Grande
Valley—History. 7. Rio Grande Valley—Environmental conditions. 8. Rio
Grande Valley—Social life and customs. I. Hall, G. Emlen, 1942– II. Black,
Mary E., 1953– III. Title.
  GF504.S685P47 2011
  363.6'1097644—dc22

                              2010044445

This material is based upon work supported by SAHRA (Sustainability of semi-Arid Hydrology and Riparian Areas) under the STC Program of the National Science Foundation, Agreement No. EAR-9876800. Any opinions, findings, and conclusions or recommendations expressed in this material are those of the authors and do not necessarily reflect the views of SAHRA or of the National Science Foundation.

Contents

Acknowledgments

This book covers a vast span of time and huge geographical area; its writing would not have been possible without the contributions of many people and organizations. The principal one of these is the Science and Technology Center for Sustainability of semi-Arid Hydrology and Riparian Areas (SAHRA) at the University of Arizona, funded by the U.S. National Science Foundation (NSF) under Agreement No. EAR-9876800. SAHRA is dedicated to promoting the sustainability of water resources in the semiarid western United States, and this work is a component of the public outreach aspect of that effort. SAHRA provided salary and logistical support without which the writing of this book would not have been possible. Fred Phillips especially thanks SAHRA's two directors, James Shuttleworth and Juan Valdés, and associate director for knowledge transfer Gary Woodard. We hasten to add that any opinions, findings, and conclusions or recommendations expressed in this material are those of the authors and do not necessarily reflect the views of SAHRA or the NSF.

Next, special thanks are due Martha Franks, a partner in the Albuquerque firm Abramowitz & Franks, and Elizabeth Hadas, our editor at the University of New Mexico Press, both of whom read the entire manuscript more than once and greatly improved it through their critical commentary and edits. Much of the graphical content of the book would not have been possible without the efforts of Matej Durcik, Mike Buffington, and Shiloe Fontes of SAHRA and Susan Delap of the New Mexico Institute of Mining and Technology.

The following individuals all made valuable suggestions or contributions to the content of the book: Steve Hansen, U.S. Bureau of Reclamation, Albuquerque; Nabil Shafike and Kevin Flanigan, New Mexico Interstate Stream Commission; Letty Belin, Office of the General Counsel, U.S. Department of the Interior, Washington, D.C.; Sabino Samuel Vigil, mayordomo, Acequia del Molino, Cundiyo, New Mexico; Anabel Gallegos, Middle Rio Grande Conservancy District; attorney Mary Humphrey, Taos, New Mexico; attorney Don Klein, Socorro, New Mexico; Elizabeth Cervantes, District I, New Mexico Office of the State Engineer, Albuquerque; Barbara Mills, University of Arizona; Robert Dello-Russo, New Mexico Office of Archaeological Studies; Julio Betancourt, U.S. Geological Survey; David Brookshire, University of New Mexico; Paul Brooks, University of Arizona; Steve Cather, David Love, Jane Love, and Richard Chamberlin, New Mexico Bureau of Geology and Mineral Resources.

For aid in searching archives and for permissions to use images, we thank Daniel Kosharek, Photo Archives, Palace of the Governors; Dean Wilkey, Archives and Special Collections, New Mexico State University Library; librarians and staff of the Center for Southwest Research, University of New Mexico; Anabel Gallegos; Steve Hansen; Robert Eveleth and Kay Brower, Socorro County Historical Society; Luann Pavlovich, San Antonio, New Mexico; and Lois Phillips, Socorro. Special contributions were made by Lisa Majkowski, Susan Delap-Heath, and Leigh Davidson, New Mexico Institute of Mining and Technology; and by Padilla's Mexican Kitchen in Albuquerque and La Pasadita Café in Socorro. Emlen Hall wishes to explicitly acknowledge the indispensable contributions of his coauthors. Without Fred Phillips, this book never would have been begun. Without Mary Black, it never would have come to an end.

Prologue
Cochiti Pueblo and a Changing River

The Rio Grande is an ancient river, the banks of which have long been home to human settlement. The small pueblo of Cochiti lies at the heart of the upper Rio Grande and its history is a microcosm of the history of the greater river. This book, which looks at the ways the Rio Grande has shaped human society in the Southwest and the ways that humans have changed the river, starts by focusing on Cochiti Pueblo.

The pueblo physically sits at the boundary between the rugged highlands of the Rio Arriba, the upper river, and the broad rift valley of the Rio Abajo, the lower river, and thus at the center of the geological events that created the river. Cochiti also resides at the interface of culture and technology. In its earliest years, seven centuries ago, the pueblo hardly interfered in the river's rhythm or flow, though it dwelt next to it. The pueblo now dwells in the shadow of the tall Cochiti Dam, the most massive type of engineered work on the river in modern times. In the intervening years, the pueblo has been both witness to and participant in the changing interactions between the river and the communities along it. A series of often contradictory values and attitudes has emerged in the form of opposing visions of the river.

These changes and contradictions reflect the ambiguity with which human society has approached the Rio Grande over the centuries. At times, and among some Rio Grande peoples, the river seemed profoundly mysterious; to others it seemed perfectly knowable. To some the river was governed by its own laws; to others it was intricately and extensively governed by the laws of man. Some felt the river had a religious integrity best left alone; others felt it invited the best and most profound technological improvements for the benefit of man. Some considered the river to be the foundation of local community welfare and sustenance that could not be separated or divided; to others, the river was the subject of individual property rights entitled to the highest and most assiduous protection. In short, to some the Rio Grande always has been the dominion of nature and to others, the dominion of man.

Is the Rio Grande a life-sustaining and seamless foundation of local community welfare, or is it a commodity to be divided up, distributed, and sold for the benefit of a broader constituency? Is it the centerpiece of an ecosystem that is valuable in its own right, or is it a supply of water to be engineered into whatever configuration will yield the maximum economic benefit to society? The boundaries between these different values are often gradational rather than abrupt, and the arguments for or against them are often subtle. The types of changes that Cochiti experienced mirror the shifting answers to these questions and the effects of those answers, unintended or not, on the Rio Grande.

The Rio Grande originated millions of years ago. It began high in the mountains of what is now Colorado and eventually pushed south through New Mexico, advancing through alternating narrow canyons, wider valleys, basins and ranges, and finally rushing all the way to the Gulf of Mexico. Cochiti Pueblo is situated just below the point where the southward-rushing river bursts out of the narrow Santa Fe canyon and opens into the middle Rio Grande valley. The geologic location forever shaped the pueblo's access to water and its relationships with neighbors who shared the same water source.

Cochiti Pueblo is one tribe that has survived for many centuries along the banks of the river and can tell the tale of change. For at least seven hundred years, this pueblo of more or less 650 residents has resided here. The pueblo's central plaza and religious kivas sit on the west bank of the Rio Grande just after it emerges from a long, narrow canyon and begins a relatively uninterrupted run through the middle Rio Grande valley, today New Mexico's most populated, most developed, and fastest growing area. For all these centuries, Cochiti Pueblo has depended on the river physically and spiritually.

Most views from Cochiti's central plaza are unchanged since the village was established. To the west, the Jemez Mountains rise ten thousand feet. East of the pueblo, the Rio Grande runs south past Cochiti. To the south, the expanse of the middle Rio Grande valley opens up all the way to the Magdalena Mountains, shimmering on a clear day seventy-five miles below. But it's just to the north of Cochiti Pueblo that the world is most dramatically defined. Looking north, the world ends and there is only sky beyond.

For most of its long existence, Cochiti Pueblo has been located at the base of La Bajada escarpment, which looms almost eight hundred feet above it. The wavering line formed by the top of the mesa divides the universe into two different riverine worlds: the confined canyons of the Rio Arriba above the escarpment, and the Rio Abajo, a wide expanse of seemingly endless valleys below.

Cochiti Pueblo in 1880, beneath the
long slope of La Bajada in background.
G.C. Bennett for Henry Brown, Courtesy
Palace of the Governors Photo Archives
(NMHM/DCA), neg. no. 13122.

But a second northern escarpment, this one man-made, has dominated
the Cochiti world for the most recent forty years, as surely as La Bajada did
before. This new division is the equally dominating, eerily uniform line now
formed by the top of the massive Cochiti Dam (see plate 1). Lying less than a
mile north of Cochiti Pueblo plaza, Cochiti Dam brought to the Rio Grande
and the pueblo the accelerated technological changes of a late twentieth-
century hydraulic society.[1] In the most concrete terms, Cochiti Dam changed
the world the Cochiti people saw, replacing the dividing line of La Bajada
with a nearer but even more imposing line of dam. More important, it
transformed the Rio Grande there and downstream, as well as the land that
was now submerged below it and the farmlands that bordered it.

According to oral tradition, the Cochiti people first came to the banks of
the Rio Grande from the highlands to the west. On their arrival, they continued their old water practices. They blocked an arroyo to divert a little water
onto the floodplain to irrigate nearby fields. They stored water for future
use behind low dams across intermittent arroyos. They mulched their fields
with gravel to help water retention. These techniques helped to provide food
for the Cochiti people. But, initially at least, the Rio Grande by which they
lived and from which they drew domestic water was within their prayers but
beyond their real control.[2]

These days the highly regulated Rio Grande looks nothing like the Rio Grande that flowed past the pueblo in pre-Spanish times. When Cochiti Pueblo was established, the river flow varied widely from day to day, month to month, season to season, and year to year. Melting snows from the northern mountains drove the river wild in the spring; torrential monsoon rains in the late summer spiked flows in July and August. Occasionally, fall storms originating in the Gulf of Mexico sat over New Mexico, drowning the Cochiti country with a week's worth of heavy rain and sending water roaring down the Rio Grande. As if in spite, nature often reversed itself and the river went dry for months.[3]

The Rio Grande accommodated these huge variations as all natural systems do. The flood plain of the river was very broad. On those rare occasions when the flows were highest, the river ran bank to bank in front of Cochiti Pueblo. However, most of the time the diminished river ran in constantly changing braided channels across the otherwise dry, thousand-yard-wide flood plain. The channels crossed and crisscrossed, offering a rich mixture of changing habitats—a pool here, an island there; sure, steady flow here; intermittent flows there.

The Cochiti people used what land they could, going so far as to plant corn on islands in the middle of the river, which allowed the crop to get moisture from seeping river water below and from above by occasional overbank flooding.[4] The earliest Cochiti residents revered the river and used it, but they did not attempt to fundamentally change it.

The Cochiti people, like other New Mexico Pueblo groups, were adaptable as well as constant in their own traditions. From the time that Juan de Oñate reached Cochiti in 1598 and Spaniards began to share the banks of the Rio Grande, the Cochiti people adopted what cultural and technological knowledge they found useful. Their Keresan language expanded to include Spanish terms such as *pés* (money) from the Spanish *peso*; *mansá·n* (apple) from *manzana*; and *siyen* (hundred) from *ciento*.[5] By the mid-eighteenth century, some Cochiti leaders spoke crude Spanish as well as their own language. And by 1776, the Cochiti Pueblo apparently had learned the basics of Iberian flood irrigation. In that year, Father Francisco Atanasio Dominguez, a Franciscan priest who had been sent to inspect and report on conditions in New Mexican pueblos, reported finding for the first time "deep wide ditches" diverting water directly from the river itself and irrigating Cochiti Pueblo lands on its banks.[6]

Indeed, most of Cochiti land and water history of the eighteenth and early nineteenth centuries involved increasing accommodation to the expanding, growing Hispanic world around it. A complicated tangle of overlapping private Spanish land claims encircled Cochiti Pueblo and its environs and squeezed the pueblo's access to resources. Some Hispanics even moved into Cochiti Pueblo.[7]

The Cochiti people reacted to eighteenth- and nineteenth-century pressures on their land resources—including control of their seeps and springs—with aggression and ingenuity.[8] The pueblo obtained a Spanish land grant of its own, which, under Hispanic law, put the pueblo at least on as firm a footing as the Spanish and Mexican land grants that surrounded it.[9] With this defensive action, the pueblo entered the world of private rights to common water.

From its place upstream at the top of the middle Rio Grande region, Cochiti Pueblo was better positioned than its Pueblo neighbors to the south for access to the river's waters. But the river itself, rather than competing claimants, posed most of the access problems for the Cochiti people and the others who lived along its banks. The primitiveness of the Puebloan and Hispanic irrigation technologies guaranteed that the Rio Grande would roll on, largely unaffected by everyone's limited use of it. By the mid-nineteenth century, the two groups may have fully used the available base flows of the Rio Grande, but the larger, wildly varying quantities of a desert river as it raged with spring runoff and the intense summer monsoon were still well beyond the reach of their engineering systems.

After 1848, when the United States took from Mexico control of the vast lands west of the Rio Grande, Pueblo land became further embroiled in a tangled web of overlapping land grants. The Cochiti people were the first Pueblos to try to apply U.S. federal protective law to their lands and Rio Grande water resources. In this they failed, finding that, in the view of courts as high as the U.S. Supreme Court, the Pueblos were considered too civilized to warrant federal protection extended to other "wandering" Indians.[10]

A more important challenge for Cochiti Pueblo by the late nineteenth century was dealing with the fact that its sacred river was a resource increasingly common to an entire regional watershed and would be affected by developments as far north as the headwaters of the river. Thus, Cochiti Pueblo was drawn into the common governance of an entire watershed. Big, basin-wide changes immediately affected the irrigated lands of Cochiti Pueblo, many of which now lay waterlogged and useless as an indirect result of massive engineering projects upstream in Colorado. Like it or not, Cochiti Pueblo was now a member of a water community in which the actions of any one group could affect other groups far downstream or upstream.

The U.S. government, which finally undertook responsibility to the Cochiti and other pueblos in the first half of the twentieth century, nevertheless was a capricious and unstable Big Brother.[11] On the one hand, it protected the pueblos entirely from the effects of any legal agreement that would be forged among the states of Colorado, New Mexico, and Texas for apportionment of the Rio Grande. On the other hand, it drew Cochiti Pueblo into a regional water authority, the Middle Rio Grande Conservancy District (MRGCD), making it a partner in sharing the federal benefits to all river users as well as the constraints.[12]

A new Cochiti diversion dam that was finished in 1932 manifested these political and technological changes on the river. This was the most permanent structure yet built on this reach of the river. For the first time, a dam reached from one side to the other of the wide and ever-changing middle Rio Grande. It could completely control river flows and diversions into the district's canals. The irrigation canals at the pueblo were also rebuilt to improve water delivery to fields, and the channel of the Rio Grande itself was reconfigured, narrowed, deepened, and regularized.[13]

A large crowd, including many on horseback, gathers on April 3, 1932, for the dedication of the MRGCD's new Cochiti Diversion. Photo courtesy of MRGCD.

The regional water district changed Cochiti irrigation and the river on which it depended, but not so fundamentally as to threaten Puebloan identity.[14] The prospects for irrigated agriculture at the pueblo improved. For the next couple of decades, Cochiti farms looked much like the other Indian and non-Indian farms up and down the middle Rio Grande: same scale, same mixture of crops, same cropping patterns, same relative success. But the Cochiti people maintained their traditional values and respect for water and all its benefits. As with other Pueblo groups, they continued to plant their blue corn alongside the more prevalent and commercial yellow corn, grown for forage, and hybrids, which produced a good annual yield.

However, the modern technological hold on the river gradually tightened as flows were constrained, the communities depending on the river grew in population, and the scope of projects lengthened to include whole reaches of the Rio Grande. For decades, federal and state water officials had searched for a site for a tall main-stem dam on the upper Rio Grande—one that could provide both storage for irrigation water in times of shortage and flood control in times of excess.

When a prospective dam site in the Taos area, deep in the northernmost canyon of the Rio Grande in New Mexico, proved physically impossible, the focus shifted to a site in the second canyon just above Cochiti Pueblo but within the boundaries of its ancestral lands. The site originally offered supplemental water for the middle Rio Grande valley at whose head Cochiti Pueblo sat. But what made the site more compelling was that it offered flood protection for the booming Albuquerque metropolis, which was hell-bent on growth. Plans proceeded for a tall, federal flood control dam at Cochiti that dwarfed the first diversion dam. Cochiti Dam drew the pueblo into the most technologically massive treatment of the Rio Grande water resource ever in its long history. Completed in 1970, the tall dam launched a thirty-year imbroglio from 1970 to the century's end. A conservative faction of the pueblo was pitted against a progressive faction that favored use of technology and the Rio Grande for its economic development potential.[15] The tall dam at Cochiti also strained relationships between the pueblo and other Rio Grande users along the length of the river.

Regis Pecos, a Cochiti leader, was clearly influenced by these modern Cochiti struggles. Pecos was born and raised at the Cochiti Pueblo in the 1950s, graduated from Princeton University in the 1970s, and did doctoral studies in history at the University of California at Berkeley. He remains a devoted member of the Cochiti Pueblo community, well connected to the pueblo's ancient history and its current rhythms and ceremonies. Pecos served several times as Cochiti Pueblo governor. On the July Feast Days, Pecos performs with the Pumpkin Clan's singers.[16]

As the mammoth Cochiti Dam rose, Pecos grew up virtually next to it. Elders told him how the U.S. Army Corps of Engineers (the Corps) persuaded a divided pueblo to accept its placement on the pueblo's property. He recalls the terror of going to sleep as a teenager while earth-moving equipment roared through the night under arc lights that bathed the pueblo in artificial light and realizing the Corps' earth-moving equipment was tearing up the burial grounds of his Cochiti ancestors.

Once water was impounded behind the huge dam, the worst predictions of dam critics proved true: the dam leaked through the fractured base on which it rested. The leakage spread downstream under the pueblo's irrigated lands and raised the water table so high there that Cochiti farming was nearly eliminated.[17]

The dam brought additional threats to the pueblo. Pushed by the Bureau of Indian Affairs and a progressive majority in the pueblo government that was devoted to economic development, the pueblo agreed to lease nearly half of its northern land base to a Southern California developer who promised to raise on the banks of the new Cochiti Lake a new city of forty thousand residents.[18] "Imagine," says Pecos, "a municipality forty times the size of the pueblo competing for control over half of the pueblo's remaining land base. Cochiti Lake would have swamped us."[19]

Cochiti Pueblo's new water reality of a tall dam split the community spiritually, technologically, and politically, and tugged it in the direction of twentieth-century American values. The recreational lake and proposed second-home development on its shores—all on the ancestral Cochiti lands—also drew Cochiti into the new realm of the passive recreational use of water for economic development. The trend became yet more pronounced when the Federal Energy Regulatory Commission (FERC) considered allowing electrical power development at Cochiti Dam. Some members of the pueblo, including Pecos, protested on religious grounds, but refused to be specific about them, despite the insistence of the commission. Divulging Cochiti religious secrets to anyone, even in private, could not be allowed. (Later, federal law would protect tribal rights to privacy on esoteric religious matters.)

In the mid-1980s things started to change for the pueblo and its stretch of the Rio Grande. The excesses of the plans for the river doomed some of them to failure. The FERC power application was withdrawn. The Cochiti Lake development went bankrupt, and the pueblo managed, with Pecos's help, to repurchase the rights of the developers. Now the pueblo itself was in the water-for-real-estate game. The pueblo also sued the Corps for damage done

Opponents of Cochiti Dam modified the advertising billboard for the Cochiti Lake development in 1971. Photo courtesy of James Bensfield.

to its lands by the leaking Cochiti Dam. To settle the suit, the U.S. government for the second time agreed to drain the pueblo's swamped lands and restore them to their farming capacity, as it had done when the MRGCD was established.[20]

The extremes of federal western water policy that prevailed in the early to mid-twentieth century, insisting that bigger is better and that the answer to technological problems is more technology, have been curtailed for now in Cochiti and elsewhere in the United States. Pecos is cautiously optimistic that the pueblo, with its adaptability and deep devotion to its own long history on the land, will survive in a way appropriate to it, a way that allows the pueblo itself to strike the correct balances between the fundamental tensions at the heart of Rio Grande history.

Today's inhabitants of Cochiti are in many ways much the same people they were seven hundred years ago; they maintain the same clan relationships, observe the same religion, and perform their ceremonies with a Catholic layer added on. They have adapted skillfully and flexibly to the massive changes around them, including changes to the natural world in which they live.

The river that flows along beside the pueblo is one of the most central and altered aspects of that changed natural world. Seven hundred years ago it was a capricious, uncontrollable entity, dry one day and, the next, a seething torrent into which no one could venture and survive. The river, alive and sacred, formed the centerpiece of the valley that was the Puebloan world. Today the river has been stripped of its wildness by the stark black ridge of Cochiti Dam. The Rio Grande that flows past Cochiti Pueblo is now much tamer and more sluggish, varying only a little in flow as the needs for irrigation change with the seasons.

Reining in the Rio Grande has brought many benefits to those who live in its valley, but the benefits have come at a price. The following chapters of this book move beyond the smaller world of Cochiti to examine the long span of river between its headwaters and where it reaches the borders of Texas and New Mexico, exploring what is gained, what is lost, and what possibilities remain for a desert river when a tide of civilization, history, and technology sweeps over it.

Roots of the Rio Grande in Deep Time

The Rio Grande was already ancient when the first humans reached its banks. Some of its rocks go back to the early stages of the history of Earth, before there were even organisms with hard shells. The course of the river was dictated by the splitting of Earth's crust in the wake of exceptionally large volcanic eruptions 20 million years ago. The productive aquifers that have sustained middle Rio Grande populations until recent times were deposited as a consequence of global cooling and glaciers in the headwaters during the past 2 million years. Even the course of the river over the last five hundred years has an elaborate history. Looking down on the Rio Grande floodplain from an airplane one sees an intricate braid of loops and whorls—the tracks of a river that was constantly changing its path.

Unquestionably, the Rio Grande has an identity that transcends human concerns. That identity begins with the building blocks of the earth itself—the physical material of which the Rio Grande basin is composed. Those fundamental earth materials have crystallized, eroded, redeposited, deformed, and uplifted over almost unimaginably vast intervals of time in order to produce the landscape and river that we see today. Today's Rio Grande is consequent on that long primeval history. The time span since the first human set eyes on the Rio Grande is infinitesimal in comparison to that long saga, yet it far exceeds the length of recorded human history.

And yet the basin's natural geographic features—the shape, contours, and composition of the land and the way its waters flow above and below the earth's surface—are what first attracted or repelled human settlers, who would in turn transform the river. Where does the rain fall? What direction does the river flow? Does it flow through a fertile plain or a precipitous gorge? What pattern does the groundwater follow as it creeps over millennia toward the river? Can good groundwater be easily sucked from the earth or is it difficult to pump or salty? The geological events of an almost unimaginably distant past provide the answers to all these questions. Geologic processes shaped the groundwater, the rivers and lakes, and the mountains and valleys of the Rio Grande basin. The availability of water is often the most important consideration determining human habitation. Unlike much of the less arid eastern United States, where outside the major urban centers the population is fairly evenly distributed, vast regions of modern New Mexico are almost uninhabited and other areas have narrow corridors of dense populations (see plate 2). In arid regions, people follow the water.

The major focus of this book is on the interaction of human populations with the Rio Grande, from prehistoric time to present day. This chapter departs from that emphasis to look at the geology, hydrology, and biology of the river system—its natural history—to provide a basis for understanding this later interaction.

Deep Time

The ancient Cochiti people felt a deep affinity for the earth upon which they lived, as do their descendants, for they believed that they had emerged into the sunlight of the present world from a *sipapu*, an opening in the ground. In truth, all people find themselves situated upon a landscape they did not create and that vastly predates them. Although the ancient Pueblo peoples possessed an acute sense of antiquity of the earth, only in the past few hundred years has modern science begun to quantify the lengths of time that have been required to shape the land of the Rio Grande and the river itself.

For most of the history of humankind our perceptions of space and time have been extrapolated from familiar human scales. The stars were not so low that they could be touched by human hand from the top of a high mountain, but perhaps they could be reached by a high-flying eagle. The mountains and hills were not created in our grandparents' time; perhaps they came into being as long as fifty generations ago. The penetrating eye of modern science has exploded these simple, commonsense scales of reality. The planets are distant from us by many thousands of multiples of the diameter of Earth, and the true stars are vastly more distant than that. Our part of the universe is arranged into a galaxy containing billions of stars, and we can see thousands of galaxies with immense stretches of space between them. The true scale of the universe is so immense that scientists have coined the name *deep space* to attempt to convey its immensity. The Rio Grande basin is only an infinitesimal part of that deep space, but it is a part.

Our perception of time is similarly limited. Fifty generations is actually less than one-half of 1 percent of the time to the origin of our species. The

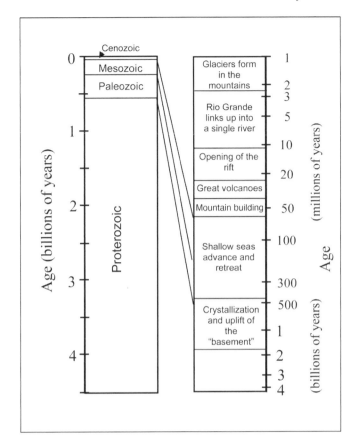

An illustration of the length of time involved in the formation of the Rio Grande basin. The time scale on the left is linear. That on the right is expanded toward the present to show the important geological events in the basin.

age of our species is less than 1 percent of the time since the extinction of the dinosaurs, but that time is only the last 14 percent of the age of Earth. The term *deep time* has come to be used to convey the scope of time over which Earth has been shaped. It is into this deep time we delve very briefly to understand the ancestry of the Rio Grande.

In the Beginning

Long before there was a Rio Grande, the earth churned and mixed deep below its crust, seas repeatedly flooded the land and retreated, the earth buckled and mountain ranges rose, volcanoes exploded and lava flowed, and the earth again cooled and settled. Over billions of years these forces shaped the land of the present Rio Grande basin. These geologic events determined the way and direction that groundwater flows, its quality, and its abundance. And they laid the path for the Great River to follow.

The age of shallow seas

The first 4 billion years of Earth (the Proterozoic; see plate 3) was a period of long, slow upheaval that carried rock and sediments deep below the earth's crust where they recrystallized into a hard, dense layer of material such as gneiss and granite.[1] Geologists refer to this hard crystalline layer, through which little groundwater can circulate, as the *basement*. By 450 million years ago, much of what is now New Mexico and Colorado had eroded down to this layer. To put 450 million years in perspective, had humans been alive then, we would be the thirteen-millionth generation of their descendants.

Next began a long period of advance and retreat of seas across the relatively flat land of this region, along with the deposition of sediment that covers much of the hard basement layer.[2] Marine animals teemed in the shallow seas and their minute shells rained down on the seabed to form limestone layers where the seas were shallow. Layers of shale were deposited from erosion of the surrounding land when the seas deepened. Layers of fine-grained beach sand were created from wave action as the coastline advanced and retreated.

The last interior seaway retreated north around 90 million years ago, but the sedimentary deposits that were left behind affect groundwater even today. Groundwater flows easily through limestone layers, supplying, for example, the extensive agriculture of the Pecos River valley. Sandy layers left in areas such as northwestern New Mexico similarly allow groundwater to easily flow and so are eagerly sought by well drillers. Shale, in contrast, obstructs the easy passage of groundwater.

In some locations, many different sedimentary layers are stacked one upon the other, creating a complex pattern of flow. In areas such as the high plains of eastern New Mexico and western Texas, extensive farmlands and numerous towns today sit atop permeable rock that contains large stores of fresh water, with the flow pushed by high water pressure. Where the geologic layers contain brackish, salty water that can ultimately be traced back to ancient seas, such as the Tularosa basin in south-central New Mexico, the groundwater flow is limited and slow and the land above is mostly empty desert.

The age of mountain building

One hundred million years ago, the movement of the plates comprising the earth's crust began to deform the area that would become the Rio Grande. A deep, underflowing layer of the oceanic floor from the Pacific Ocean, which had worked its way beneath North America, dragged against the continent above it, causing the land surface to buckle and rise. The continent thickened and eventually rose far above sea level, where it has remained to the present day.[3] Many of the high areas that formed this way now constitute parts of the Rocky Mountains.

The mountains rose well above their surrounding lowlands. Basins sank between the mountain ranges and collected older sediments that eroded from these higher elevations. This evolution of basins and ranges was critical to the formation of the Rio Grande. The high ranges catch the atmospheric moisture that moves east across North America, providing the rain and snow needed to sustain a large river.

The age of supervolcanoes

The next big period of geologic change left its mark on the land in similarly dramatic fashion: through volcanic explosions that left ash and pumice deposits that still affect runoff and streamflow.

What set off these explosions? The long episode of mountain uplift ended around 40 million years ago when the two plates underlying the Pacific Ocean and North America shifted direction and started to slide past each other in a north-south orientation.[4] The change in motion sliced off the slab of ocean floor that was being shoved eastward below the continent, causing it to sink into the earth's mantle, and forced hotter rock up from below, which warmed the continental crust. Rock near the base of the continent melted and rose in the fashion of the colored blobs in a lava lamp. Gases accumulated near the tops of these magma blobs and eventually burst forth. The molten rock expanded as pressure was released. The effect is much like what happens when a bottle of champagne is shaken and the cork popped: a column of froth shoots skyward.

Forty or fifty of these immense pops, sometimes called *supervolcanoes*,[5] devastated the landscape of what is present-day New Mexico and Colorado 37 million to 21 million years ago (see plate 4). The biggest of these produced more than one thousand cubic miles of lava—enough to completely fill Lake Erie—and are among the largest eruptions in the entire history of Earth.[6] They left circular holes in the earth as large as fifty miles across.

Thick layers of pumice blanketed the entire landscape, often followed by incandescent clouds of volcanic ash that settled and cooled into hard volcanic rock that transmits water very well. In areas that get less rainfall, such as western New Mexico, streams and rivers are rare, because the precipitation sinks easily into the volcanic rocks and down into the groundwater, instead of running off in surface flows.

The beginning of the big river rift

The pulse of heat produced by the sinking oceanic slab dissipated 25 to 20 million years ago, and volcanic activity waned. This period of calm set the stage for the birth of the Rio Grande. In what is now the Rio Grande basin, the earth's crust was left vulnerable, made thicker than surrounding areas from the compression of land that formed the basins and mountain ranges and also softer from the subsequent heating. As the volcanoes ceased, the elevated crust divided and very slowly flowed east and west along a north/south fault zone,[7] creating a geological *rift* that began to open and define the boundary of the Rio Grande drainage basin. Were it not for the rift, the streams that originate in the northern mountains would not collect today into a single large river, but would spread out in many directions.

That the Rio Grande follows a remarkably linear north/south course from its headwaters in the San Juan Mountains to El Paso, Texas, is no accident. The river follows the depression left by the opening of the rift. The rate at which the rift opened was greatest 20 to 10 million years ago, but it is still very slowly growing wider. It opened because the land was stretching and extending, its center was subsiding, and the elevated blocks along its margins slumped into the depression. It collapsed in stages, following a complicated geometry, with some areas dropping deeper than others and sections opening at different times, but generally from north to south through time.

The very formation of the rift ensured that it would eventually be filled. The wettest portion of the Rio Grande drainage basin is in the highlands around its headwaters. As the climate became moister due to increasing elevation of the uplifted areas, increasingly vigorous northern streams began to bring abundant coarse sediment into the subsiding streams. This sediment filled the basins faster than they were sinking; consequently, the basins filled to the brim. Streams began to spill from one basin to the next downstream. The rift now became a kind of trough or drain that channeled runoff into a single path. The more water that flowed into the rift, the more sediment the river carried, and the quicker it filled basins downstream. Each basin that was filled added its flow to the elongating river.

Once the river filled a basin to the brim and spilled to the next one, the flow generated an elongated, sinuous zone near the center of the rift. This zone filled with clean, coarse sand and gravel, which delineates the bed of the ancient river. Large amounts of good-quality water can be found in these ancient channel deposits. Toward the edges of the main streambed, in the floodplain of the ancient river, the deposits are thinly layered, and the sediment contains finer silts and clays. Both the history of the formation of the

Rio Grande rift and the likely location of good quality water can be determined by looking at the types of sediments in the Rio Grande floodplain.

Around 10 million years ago, the Rio Chama met up with the Rio Grande and its rift near Espanola, New Mexico, forming a V-shape and pushing the combined strength of the river and tributary ever downhill and southward. By 7 million years ago, the river penetrated the northern basin around present-day Albuquerque. Around 5 million years ago, it reached southern New Mexico. It was "only" about 2.5 million years ago that the northern Rio Grande breached the final barrier and joined its southern portion, which flows to the Gulf of Mexico. This southern stretch had previously drawn most of its runoff from the Rio Conchos in northern Mexico.[8]

Strangely, the sub-basin to most recently link up with the main river is the northernmost one. Up until about half a million years ago, the southern portion of the San Luis basin in Colorado was a vast lake, into which what is now the main stem of the Rio Grande dumped.[9] Finally the lake overflowed the sill at its southern end and flowed south to join up with the rest of the river. As it did so, it cut the dramatic Taos Gorge, beloved of river rafters today. From the point that the Rio Grande flows out of the San Luis basin, the origin of the basin as a drained lake is so obvious that it was commented on by early explorers.[10]

Much of the water presently used in the Rio Grande basin is actually groundwater, which is available largely because of the geology of the Rio Grande rift. The groundwater feeds the base flow of the river—the flow that is not dependent on runoff from bursts of snowmelt in spring thaws or summer cloudbursts.

Deconstructing the Present Landscape

The Rio Grande basin vaguely resembles a giant seahorse (see plate 2), a shape produced by its geological history. The eastern part of the basin is bounded nearly everywhere by the uplifted areas that define the eastern edge of the geological rift. These uplands include the Sangre de Cristo Mountains in the north, and the Sandia, Manzano, San Andreas, and Organ mountains farther south. Most of these mountain ranges are narrow spined, formed of the hard, dense basement rock that was uplifted and eventually exposed by erosion. In the northern portion of this eastern side of the basin, many of the peaks are quite high: several in the Sangre de Cristo range extend more than fourteen thousand feet. Farther south, the boundary mountain ranges diminish in height to eight thousand feet or less, occupy a relatively small area, and so contribute little water to the Rio Grande drainage.

An aerial view of the Rio Grande gorge near Taos shows the canyon cut when the San Luis basin linked up with the rest of the Rio Grande about five hundred thousand years ago. Courtesy Palace of the Governors Photo Archives (NMHM/DCA), neg. no. 155878.

It is in the northwest where the great majority of Rio Grande flow originates. Here, the basin boundary is formed by the massive San Juan Mountains (several peaks reach up more than fourteen thousand feet), the highlands around Rio Chama, and the Jemez Mountains. Unlike the eastern ranges, these mountains are formed of more easily eroded volcanic rock, and so they contain many large, high-elevation basins that collect precipitation. The San Juan and Jemez mountains are also well positioned to intercept the winter storms that move across the continent, and they accumulate thick snowpacks during the winter. As the snow melts rapidly in the spring, water flows into the deep canyons in the interior of these ranges and then is funneled down the Rio Grande. Much of the snowmelt also sinks down through the porous volcanic rock into the groundwater and is later discharged to the Rio Grande downstream, moderating the snowmelt pulse.

The mountains that form the southwest edge of the rift—the Magdalenas, San Mateo, and Black ranges—cover a smaller area and are relatively dry, so they contribute little flow to the Rio Grande even though they reach up to eleven thousand feet in elevation. Similarly, only moderate flow is added to the Rio Grande from the confluence of two streams from west-central New Mexico—the Rio Puerco and the Rio Salado. These rivers broke through the ancient wall of the rift to join the Rio Grande, but they drain a relatively dry region with few areas of high elevation.

Rainfall and the River

The Rio Grande is just a tiny drop in the global hydrological system, a cycle in which water vapor is transported from the oceans over the continents in vaporous clouds, changes to a liquid (rain) or a solid (snow and hail) and falls to earth, runs off, infiltrates, evaporates, and ultimately returns to the oceans. In the Rio Grande basin, as everywhere, this system depends on a complex interaction of features and processes, from the global movement of water vapor, to local idiosyncrasies such as elevation, rock type, and vegetation within the basin.

The average yearly precipitation throughout the basin varies considerably, due mainly to two factors.[11] One is global atmospheric circulation, which dictates that most winter storms move from the Pacific Ocean across the northern and central sections of the continent. Storms are less and less frequent farther south. The other main factor is elevation: as air masses ascend the high mountain ranges, the air cools and water vapor condenses, producing rain or snow. For every thousand feet in elevation increase, temperatures drop about three degrees Fahrenheit, and annual precipitation increases about three inches (see plate 5).[12]

The San Juan Mountains, where the Rio Grande has its headwaters, enjoy a happy confluence of these factors. They are located where storms track especially frequently and also have a large area at quite high elevation. Consequently, they receive the most precipitation in the Rio Grande basin. The Sangre de Cristo Mountains, which are equally high but receive most storms only after they have passed over and expended their moisture on the San Juans, get much less rain and snow. Although fairly high, the southern mountains, such as the San Andreas and Black ranges, receive much less precipitation than the San Juans because they are south of the major storm tracks.

Winter storms are a particularly important source of Rio Grande water. This is partly because the mountainous northern area of the basin gets most of its precipitation in the winter. Another reason is that less moisture evaporates at colder temperatures, so it remains available to run off when the snow melts in the spring.

In contrast, much of the summer rainfall is sucked up by plants and returned to the atmosphere, with less left to run off.[13] Summer storms also follow a different path. In summer, the interior of North America heats up and draws in moisture from the Gulf of Mexico and from the Pacific Ocean off western Mexico, as compared to winter storms, which often arc down from the northern Pacific. This summer moisture generally does not move in the form of storm fronts, but infiltrates northward. Afternoon heating of the land causes hot, moist air to rise and create thunderstorms. This North American monsoon pattern is strongest at the southern fringe of the Rio Grande basin and dies out to the north.

The Plant Factor

Throughout the Rio Grande basin, native plant life varies dramatically with elevation and rainfall (see plate 6). In the warmer, more arid regions, the growth of individual plants is limited mainly by the amount of water they can take from the soil. Those that are best adapted to this climate limit the water they lose through transpiration by keeping their stomata—the microscopic pores on their leaves or stems—open less than do plants in wetter regions. The price they pay is growth: the new tissue they can produce is severely limited. This is the fundamental reason deserts are dotted with only a sparse layer of straggly plants, while tropical regions with abundant precipitation are densely covered by rainforest.

Vegetation affects the water supply as well as being affected by it. Plants are a crucial component of the Rio Grande basin's water balance. Plant characteristics such as type and density largely control the amount of water that runs off the land after rain or snow, the portion that evaporates from the soil, as well as the amount that is sucked up in root systems and breathed back out to the atmosphere. At the ground level, a good carpet of plants, plant roots, and plant litter encourages storm water to soak into the ground and keeps the soil intact. The plant canopy above can also affect the water balance, particularly in areas that get winter snow, with less dense canopy allowing snow accumulation to increase soil moisture and very dense canopy losing up to half its snow water to sublimation—where the snow that hangs upon the needles goes straight from solid state to water vapor and is lost to the atmosphere.[1]

Streamflow also depends on the vegetation cover within the stream's drainage basin. In arid portions of the drainage, the vegetation uses nearly all of the unevaporated water. Plant growth and viability is very dependent on water availability and precipitation. Little water is left over to run off or to infiltrate the land surface and recharge the groundwater.[2] The relatively small fraction of precipitation that does infiltrate through the soil flows slowly beneath the ground to emerge eventually into the beds of streams. This fraction supplies the flow of small streams that ultimately join to form the Rio Grande. In contrast, in areas with considerable rain or snow, plant growth is usually limited by other factors, such as competition for sunlight or essential nutrients. Excess water runs off or infiltrates into the groundwater, supplying large streams and rivers.

Big changes in vegetation cover are reflected in the basin hydrology. Deforestation of the land, such as from logging and grazing in the early 1900s or from massive plant die-offs from beetle infestations in western piñon-juniper and lodgepole pine forests in the 2000s,[3] upsets a basin's entire water balance. Soil that has been disturbed or denuded encourages rapid runoff, raising the height of flood peaks. The soil is prone to being stripped of nutrients, carried off with overland flows, and dropped out on the beds of the rivers, deteriorating the water quality and further increasing the risk of flooding.[4] A river is intimately linked to the vegetation that covers its drainage basin.

Notes
1. Veatch 2008.
2. Sandvig and Phillips 2006.
3. Breshears et al. 2005.
4. Cooperrider and Hendricks 1937.

Accelerated Change

Geology, climate, and ecosystems are never static, and the Rio Grande basin is certainly no exception. However, change that used to come slowly, over eons, has clearly accelerated in the years since humans entered the basin, increasingly so in modern years. Flows in the Rio Grande tell part of this story (see plates 7 and 8). Modern flows in the Rio Grande—with respect to both season and location on the river—only mildly resemble the patterns before development.

In this and in many other respects, humans have profoundly impacted the river, particularly since the late 1800s. In future chapters, we will explore why and how humans have so indelibly changed the behavior of the river and the often-unexpected outcomes that have resulted from this interference.

Early Cultures

TABLITA RAIN SONG, COCHITI[1]
Now come rain! Now come rain!
Fall upon the mountain; sink into the ground.
By and by the springs are made
Deep beneath the hills.
There they hide and thence they come,
Out into the light; down into the stream.

Look to the hills! Look to the hills!
The clouds are hanging there,
They will not come away;
But look, look again. In time they will come to us
And spread over all the Pueblo.

Look at us! Look at us!
Notice our endurance!
Watch our steps and time and grace,
Look at us! Look at us!

Just thirteen thousand years ago—the blink of an eye in geologic time—the land of present-day New Mexico was solely the realm of the beasts. Then humans started leaving their tracks in the Rio Grande Basin, and the long relationship of water and people in present-day New Mexico began. People in large numbers would eventually settle and farm near the rivers, particularly the Rio Grande and its tributaries, but this process took millennia to play out. The connection between early human populations and the Rio Grande was tenuous and continued as such until relatively modern times; until after AD 1100, this area saw only intermittent human settlement.

By the time Juan de Oñate reached the Rio Grande in 1598, eighty-one good-sized pueblos were occupied on the Rio Grande by an estimated one hundred thousand Puebloan people. The Pueblo tribes would suffer terribly in the centuries following the arrival of the Spaniards. By the time of the Pueblo Revolt in 1680, when the Pueblo groups collectively but temporarily expelled the Spaniards from the Rio Grande and other western pueblos, the number of pueblos was down to just thirty-one, with a total population of around fifteen thousand. Most of these pueblos were abandoned between 1600 and 1640, due to the ravages of epidemics unwittingly brought to the New World, forced labor, raids by non-Puebloan tribes facilitated by the Spaniards' introduction of the horse, loss of land and resources, and drought and famine. When Spanish rule was restored in 1692, only twenty pueblos remained or were resettled.[2] Of these, some eighteen pueblos remain today, a testament to the tribes' resilience and their determined commitment to a sustainable, agricultural way of life. This lifestyle remains firmly rooted in a deep and enduring philosophical and spiritual appreciation of the connections of land, life, and water.

In the thousands of years between the first appearance of humans in the Southwest and their eventual settlement in substantial numbers along the Rio Grande and its tributaries, the human/water connection was closely interwoven with climate fluctuations and the fits and starts of technological advances achieved through trial and error. Hunting as a primary means of survival gave way to hunting and increasingly sophisticated gathering of food and plant materials, followed by the revolutionary idea of farming. Tending to crops required reliable access to sustainable sources of water, tethering people to geographic areas suited for farming, and required attention to the careful timing and placement of crops. As the number of people that could be sustained through farming and technological advancement increased, so did the risk of over-reliance on resources to support the larger populations in concentrated numbers and the risk of unintended consequences when technological experiments failed.

Earliest Immigrants

The very earliest people to reach the Southwest, around twelve thousand years ago, were Paleoindians or Paleoamericans. They were distinguished from later cultures by their heavy reliance on big-game hunting. They came at the end of the Pleistocene, when the last major glaciers were retreating in North America, and were drawn into a world inhabited by huge mammals, or *megafauna*: mammoths, mastodons, camels, lions, saber-tooth tigers, and musk oxen. Lakes dotted the landscape at that time, and people were attracted to the lakes and rivers in large part because their prey gathered there to drink. The animals were easy, plentiful targets. A catastrophic event around 10,900 BC—variously and controversially attributed to the explosion of a comet on or above North America, or overhunting by the Paleoindians themselves—suddenly wiped out the megafauna.[3] This extinction coincided with advances in hunting technologies and regional drought and was followed by sharply cooler weather.[4] Only the smaller mammals that today we consider "big" game—such as bison, deer, elk, antelope, bear, and bighorn sheep—were left on the continent.

By 7500 BC the Cody people—one of the last of the Paleoindian cultures—had established hunting sites along the Rio Grande, Rio Puerco, San Juan, and other streams and canyons. They were expert enough hunters to slaughter two hundred bison at a time, assembly-line style. But over the next two thousand years, the climate entered the Altithermal (high-temperature) period, becoming drier and even hotter than it is today. The lush grasslands dried up, and most buffalo herds no longer roamed as far south as they had, remaining even during the winters in what is now Colorado, Nebraska, and Wyoming.[5] The people adapted by shifting to nomadic hunting and gathering. People would eventually return to cluster their settlements along the Rio Grande and its tributaries but not in their present locations until around AD 1300.

The next cultural group to move through the Southwest—the Archaic people—reached the Rio Grande area from the west around 5500 BC and made full use of the natural resources available to them as hunters and gatherers, becoming particularly good at exploiting a variety of plants.[6] The grip of the high temperatures of the Altithermal eased during the Late Archaic period, around 3500 BC. Hunting and gathering became easier to accomplish, so human populations expanded.

The most enduring achievement of these Late Archaic peoples was the adoption of domesticated agriculture: they learned to sow seed, nurture plants, and preserve the seed of plants best adapted to their region. They became farmers and tended corn, beans, and squash. With the transformation to an agricultural way of life, sustaining crops with water became the most important means of sustaining the people. Ensuring water for crops became and remains, perhaps, the most culturally cherished enterprise and mission of native peoples in the Southwest, ultimately leading the tribes to the Rio Grande and its tributaries.

Mother Corn

Maize first appears in the archaeological record of the Southwest between 2100 and 1400 BC at a number of sites in southern Arizona and western New

Mexico; it subsequently became firmly established throughout modern-day New Mexico and southern Arizona.[7]

It took more than one thousand years, however, for the Late Archaic peoples to make corn an established part of their diet, as they gradually learned where and under which conditions it would grow.[8] Selection of the best seed corn over the centuries also likely gradually improved the maize's productivity. The earliest agriculture was accomplished simply by planting in areas that were suitable to a particular crop, where rainfall, runoff, or groundwater were sufficient for a harvest. People adapted to the plant, studying it carefully to discover how they could help it flourish. These early farmers took the rain that nature provided and built their lives around helping the plants capture that water. They didn't need rivers to irrigate their crops, only the knowledge gained by closely observing where the corn grew best.

Over time these early farmers discovered what we know to be true today: maize grows best on land that gets fifteen to twenty inches of rain per year and has around a 120-day growing season, the period between the last killing frost of spring and the first killing frost of fall. With a longer growing season, less rain is needed; with more rain, a shorter growing season is sufficient.

Agriculture transformed these Late Archaic peoples culturally, nutritionally, and economically. Although corn became the most important plant food in the Southwest by AD 500, many other plants suitable for food and fabric helped in this transformation.[9] It was long thought that people brought squash to the Southwest around 800 BC, beans a little later, and cotton around AD 400, but these plants may have been introduced together as early as around 1000 BC.[10] The important fact is that together corn, beans, and squash constituted the golden triangle, providing a nearly complete human diet.[11]

The success of this agricultural transformation led to more dependence on farming. As crops were successfully harvested, populations expanded. Human populations became dense enough on the landscape that hunting and gathering could no longer meet human nutritional needs. Populations became dependent on farming to reduce shortage risks, and corn became a necessity rather than a supplemental food.[12] Instead of relying on hunting and gathering to get through lean times, the emphasis shifted to building up multiyear stores of grain and seed corn.

How to Water a Corn Crop

In the period following the Archaic to AD 1000, three principal cultures arose throughout the Southwest—Anasazi/Puebloan, Mogollon, and Hohokam—each with a distinct cultural fingerprint in such practices as the way they built their houses, crafted and decorated their pottery, and grew their crops (see plate 9). It was a time of grand experimentation and learning. Farming evolved to suit an area's particular climate, geography, and resources, as well as the farmers' level of technological development and even, perhaps, their philosophies. The techniques they used also varied widely in the degree of control they exerted over water resources. Although each cultural complex has been traditionally associated with a special technique for getting water to plants or their plants to water, all of them actually used a remarkable variety of techniques to achieve their harvest.[13] The classification of these cultures by archaeologists, however, is a moving target, with recent work using the label *early agricultural complex* to describe the very early development of irrigated agriculture in southeast Arizona.

Early Anasazi/Pueblo cultures (ancestors of those who now populate the modern Rio Grande pueblos) and the Mogollon (who would disappear as a distinct cultural group between AD 1000 and 1500) used dryland farming, or precipitation-dependent farming, as one way of matching water to corn crop. Simple in concept, but not in practice, the successful dryland farmer selected his field location carefully to improve the chances of intercepting runoff from rain, planted in a location that was seed-friendly in terms of soil moisture, and timed the sowing of seed carefully. The spotty southwestern rains also had to cooperate and fall at the right place and in the right time of the growing cycle. Since dryland farmers relied only on rainfall to water their crops, they generally had better results in higher elevations where more rain falls. This farming method is still practiced by a few modern Puebloan groups, including the Hopi, Zuni, and Acoma, who plant in a wide range of settings, knowing that some places will get rain and others will not, thus spreading the risk.

A second watering technique manages the runoff from rain and snow more actively and is referred to as *runoff irrigation, floodwater farming*, or *ak-chin farming*.[14] The Mimbres Mogollon—a Mogollon cultural branch in what is now southwestern New Mexico—used this method, as did Pueblo cultures and others throughout the Southwest. They planted crops in canyon bottoms that had intermittent streams, in order to exploit the high water table, or on alluvial fans or contoured terraces that received natural overflow. Spring planting allowed the plants to draw moisture from the soil early in the growing season until snow could melt and the runoff could reach them. The farmers made minor structural improvements to manage the water, such as temporary check dams of logs, brush, mud, and stone to help slow and retain the flows.[15] Bordered gardens, waffle gardens, and contoured terraces helped retain runoff water and prevent soil and nutrient loss.

Waffle gardens have square planting areas surrounded by low berms, which catch the rainfall and deter loss of soil. In photo, waffle gardens are being planted near Zuni Pueblo around 1910. Jesse Nusbaum, Courtesy Palace of the Governors Photo Archives (NMHM/DCA), neg. no. 043170.

A third type of agricultural system used by the ancient peoples of the Southwest—collecting and diverting water from perennial sources by means of ditches or canals—is what many might consider true irrigation.[16] This includes diversion dams, canals, headgates, wells, and reservoirs. The Hohokam of central Arizona adopted and mastered the practice of canal irrigation. They succeeded at irrigated farming because they were able to rely on comparatively constant water sources such as the Salt, Gila, and Santa Cruz rivers, and enjoyed growing seasons that were longer than those of either the Mogollon or Anasazi/Pueblo. Archaeologists have now found evidence of irrigation canals in southeastern Arizona that date back to 1250 BC. These show that the early diversions were built to harvest floodwaters, but they became increasingly sophisticated structures, until, ultimately, they were designed to capture the flow from perennial streams.[17]

The Hohokam refined the practice of diverting and conveying water to an impressive degree, building main canals from the Salt and Gila rivers that were up to seventy-five feet wide at the top and twenty-five miles long. Canal building was not limited to the Hohokam, however, and ancient canals continue to be discovered in other areas of the Southwest. Early irrigation systems with small canals were recently discovered in the Zuni region of New Mexico in two sites that included fourteen canals dating between 1000 BC and AD 1000.[18]

But massive systems such as those built in the Salt River valley often relied on a fragile ecosystem. As remains true to this day, significant advances that lead societies to depend substantially on a single technology or resource put whole populations at risk if those systems fail. The Hohokam lesson is critical for modern residents of the Southwest to consider, depending as they do on streamflow of just a few rivers (notably the Colorado and Rio Grande), while simultaneously draining the millennia-old groundwaters to sustain their exponentially greater numbers and demands. Fifteen hundred years of Hohokam occupation along the Salt River apparently came to an end quickly when the immense canal system became overstressed by burgeoning populations, social fragmentation, and climate change, a situation that has clear parallels with the current-day situation along the Rio Grande and throughout the Southwest. Populations that had become dependent on the canal system to water their crops could not cope with the extreme fluctuations from flood to drought seen in the Salt River area from around AD 1356 to 1384 and along the Gila River from AD 1420 to 1428. After centuries of relatively reliable and predictable river flows, these weather extremes overwhelmed the Hohokam farmers, and the systems broke down. For example, the Hohokam groups in the present-day Phoenix area seem to have been done in by two years of extreme flooding, followed by six years of drought.[19] The Hohokam eventually disappeared altogether as a distinct culture by the mid-fifteenth century, the people abandoning their homelands and becoming absorbed elsewhere.

One might think that the Rio Grande, farther east, would have been an easy target for early farmers to use as a water source for direct irrigation diversions, but no one attempted this. Using the main stem of the Rio Grande as an irrigation source seems to have posed more massive problems than diverting the waters of the Salt, Gila, or Santa Cruz rivers, including the danger of being wiped out by floods, the difficulty of harnessing the deep and rapid flows, the density of the riparian vegetation, and the potential for disease and insects associated with stagnant water where the floodplain stretches out.[20]

While the Hohokam were perfecting and practicing canal irrigation over hundreds of years, the Anasazi and their Puebloan ancestors who farmed northwestern New Mexico and would eventually settle the modern Rio Grande pueblos were becoming primarily floodwater farming specialists.[21] What they lacked in terms of a sustainable water supply for their crops they made up for in resourcefulness.

By the Pueblo II period (AD 900–1100), farmers of the San Juan basin were working nearly every possible cropland, occupying small settlements so that if crops at one location failed they could trade with another settlement that received better rain that year.[22] With some important exceptions, most pre-Puebloan settlers from the eighth to twelfth centuries relied on collecting runoff to water their small garden plots instead of using stream diversion and conveyance systems.[23] To catch moisture and retain soil, they made bordered gardens with retaining walls, gravel-mulched gardens, contoured terraces, and check dams to slow the flow in streams that ran intermittently.[24]

The Chacoan Exception

One of the few locations where the residents practiced all methods of water control—including irrigation from permanent sources—was at Chaco Canyon in northwestern New Mexico (west of the Rio Grande), birthplace of a remarkable social complex referred to by some as the *Chaco Phenomenon*. Chaco culture has been called "the most grand and most complex society in all of prehistoric North America,"[25] the peak of which spanned the years from around AD 1020 to 1125.[26] Here, the Anasazi inhabitants built an agricultural system that included diversion dams, canals, and headgates, as well as all other less complex or invasive techniques of early Puebloan agriculture.[27]

Aerial view of excavation of Pueblo Bonito, Chaco Canyon National Historic Park, taken in 1941. Harvey Caplin, Courtesy Palace of the Governors Photo Archives (NMHM/DCA), neg. no. 58337.

Chacoans guided runoff from mesa tops to gridded fields in all twenty-eight tributary canyons north of Chaco Wash, but they did not directly divert the flow in the main wash.[28] They adjusted their methods as conditions in the channels changed from *aggradation* (the building up of sediment and raising of the water course, which increased the chances of breaching and flooding) to *entrenchment* (the lowering of the water course from erosion, as by arroyo cutting). For example, in the early 1000s, some evidence suggests that they patched and rebuilt with masonry a natural sand-dune dam that had been breached around AD 900.[29] More aggradation followed. (Similar problems would be repeated along the Rio Grande in the twentieth century, following other efforts to control the river.) By the late eleventh century, the Chacoan Anasazi mainly returned to an earlier ak-chin type of farming.

The Chaco Phenomenon has been extensively documented, discussed, and debated. Its area of influence is now thought to have extended over at least eighty thousand square miles in the Southwest. It included exquisitely constructed sandstone great houses, an extensive road system, and connecting signaling stations (see plate 10).[30] But the culture could not sustain itself. By AD 1130, the complex collapsed. The residents abandoned the big houses, departed the Chaco region, and moved north toward the San Juan River and Mesa Verde.

Many factors are thought to have contributed to the failure of the great Chacoan society.[31] These include unsustainable agricultural practices[32] and a return to overinvestment in one primary subsistence method—capture of runoff—that could not sustain the population during drought.[33] Jared Diamond concludes that the ultimate cause was "living increasingly close to the margin of what the environment could support."[34] While populations were swelling, increased ritual demands and an appetite for imported luxury goods widened the gap between the haves and have-nots. Those who lived in the big houses enjoyed a disproportionate amount of the society's food, resources, and goods, while those in the farming communities suffered from starvation, anemia, and arthritis from stooped labor.[35] The child mortality statistics paint a grim picture: children younger than five accounted for 26 percent of burials within the more prosperous inner Chaco complex, while the rate rose to 45 percent in the poorer outlying areas of the Chaco system.[36]

To explain the fall of Chaco, some modern Puebloan people point to a need to simplify and return to egalitarian ideals. Paul Pino, an official and member of Laguna Pueblo tribe says, "At Chaco there were very powerful people, who had a lot of spiritual power, and these people probably used their power in ways that caused things to change. And . . . *that* may have been one of the reasons why the migrations were set to start again . . . because these people were causing changes that were never meant to occur." Edmund Ladd, a Zuni archaeologist, suggests that there may have been a conscious decision by tribal groups not to continue the practices that exerted power over natural forces, as was perceived to be the case at Chaco.[37]

Where did the Chacoan people go? They migrated from the heartland of ancestral Pueblo territory to areas at higher elevations. Areas that were still occupied after Chaco Canyon occupation ended include the upland areas that ring the San Juan basin: the middle San Juan River area, the Chuskan slope, and the Puerco and Zuni rivers to the south. One might think the move would have been downhill to the bigger rivers, but in fact the movement was uphill, to the promise of more natural moisture in the uplands. In some respects, they went back to basics. They went to where the water falls naturally and with more frequency, and where they had more options. They reverted initially to the uplands north, south, and west of the San Juan Mountains, including Mesa Verde.

But these resources were also the source of intense and bitter competition. Thirteenth-century evidence of murder, pillaging, and malnutrition reveals a period of desperation and strife. Around AD 1230 Puebloan groups began to regroup in fortified, defensible cliff houses and mesa-top sites such as Mesa Verde. Between 1270 and 1300 even these structures were abandoned. Erosion, arroyo cutting, warfare, factionalism, disease, overpopulation, and overexploitation of the resource base have all been suggested as contributing to the move.[38] Severe drought in the San Juan drainage from 1276 to 1299 remains probably the main factor for the desertion of thirty thousand Anasazi from the San Juan valley, leaving the area for others to settle.[39] That these mass migrations tended to coincide with megadroughts suggests an overdependence on maize agriculture in an area only marginally suited to it, which left the populations particularly vulnerable to climate fluctuation—a recipe for disaster.[40] Each hypothesized cause has its critics; more likely the causes were multiple, complex, and interrelated. The fact was that life became unbearably difficult in the region because of the competition for food, violent raids, drought, diminished crops, and resultant starvation.

Archaeologist Linda Cordell makes a compelling case for a domino social effect as more and more individuals felt the need to leave:

> "What can it have been like to watch kin and friends depart, knowing that there would be fewer people to help with the daily work, fewer visitors bringing news, and not enough people to hold the important ceremonies that marked the passage of lives or the change of seasons? Surely the bonds of human interaction, fellowship, and secure social life outweighed any advantages in staying."[41]

Final Destination: Rio Grande

Many of the new immigrants into the Rio Grande region at first spread widely through the upland regions, but by the 1400s, the resources of these mountainous areas couldn't support the expanding populations, and settlements moved downhill—southeast—to areas with good and reliable water. Farmers planted corn in the floodplains, where the water table was higher, and on terraces along the Rio Grande and its tributaries. Gradually, people gathered into larger communities, and they formed modern pueblos.[42] This shift to larger villages gave them improved security, access to perennial sources of water, and a larger labor pool.[43] The move to lower elevations and to permanent water resources was substantial and long lasting, at least until the Spaniards arrived and many of the pueblos were abandoned.[44]

While the Rio Grande basin provided more access to water that flowed year-round, the movement to the river was not with the aim of diverting river flows to irrigate crops. The tribes that settled on the northern Rio Grande or its tributaries seem to have practiced little if any irrigation by canal or ditch from the rivers until after the Spanish contact.[45]

By 1400 the Rio Grande, Rio San Jose, Jemez River, Rio Puerco, Rio Chama, Zuni River, Rio Pescado, and other larger drainages in the area were home to around thirty major villages. These numbers swelled further over the next 150 years. The settlements were most closely spaced along the Rio Grande and its tributaries between Taos and San Marcial.[46] Tributary farming in these dispersed settlements meant that each tribe had access not only to a relatively reliable source of flowing water, but also to the uplands surrounding it. These uplands crossed several ecological zones and so provided each tribe access to diverse resources and land suited to a variety of crops.

When the Spaniards reached the Rio Grande in 1540 and found the many good-sized pueblos that lined the banks of the Rio Grande and its tributaries, they saw farmers but not irrigators. The Spaniards were keen observers of agricultural practices and carefully documented any irrigation being practiced in the Sonoran region, to the south and west. But irrigation by the Puebloan groups was not noted by the early Spanish, except in very isolated and specific circumstances.[47] All early Puebloan Rio Grande farmers did, however, use some form of water control, principally floodwater farming, in addition to dryland farming, as their main means of production.[48] This was sufficient to sustain their crops and way of life, and both techniques were well suited to growing corn.

Before Spanish contact, human impacts on the river system were relatively small and short lived. Arroyo-cutting may have occurred in some isolated areas as the result of humans manipulating the flow, but the river always prevailed and was undiminished. The Spanish and Mexican farmers who followed demanded more of the river and taught the Puebloans to make similar demands. But none of these groups conceived of controlling and subjugating the mighty Rio Grande.

More Than Just a Resource

Water was obviously critically important to the agricultural successes of the precontact cultures who settled the Rio Grande lands and its tributaries, as it is to the surviving Pueblo people and to farmers all over the world. The changes that people wrought in prehistoric times are revealed through archaeology—the path of clues left behind from their activities and technological refinements. But knowing the full scope of these ancient peoples' beliefs about and attitudes toward water is impossible. They left no written records, and we can only infer what their attitudes might have been and how those attitudes might have influenced their actions toward their surrounding environment.

What their modern descendants say about water provides our best clue to the ancient thinking. Pueblo people have been persistent and faithful guardians of oral tradition over the centuries, and tribal lore consistently acknowledges the importance of water as a gift of the creator, indivisible from a sustainable and enduring traditional way of life. Water is to be blessed and cherished, certainly not taken for granted, let alone exploited. It is a revered part of the abundance of a natural world that contains—rather than works against—humanity. This is exemplified in the beliefs of the Hopi, who, as a consequence of their geographical isolation, may have most consistently maintained the social, religious, and subsistence patterns of their proto-Puebloan ancestors.[49]

The practice of *katsina* (kachina) ceremonies emphasizes the importance Puebloan people gave to water. The ceremonies developed in the Little Colorado area around 1300 and spread quickly across the Pueblo world; they are still practiced at Hopi and Zuni pueblos. Katsinas are embodiments of many aspects of nature, religious figures, plants, animals, and even cultural groups. Pueblo oral tradition, anthropologists, and archaeologists have all suggested that the katsina societies represented a new social order that offered an alternative to dilute the power of individual families and clans, and the social stratification that apparently was rife at locations such as Chaco. Above all, katsina dances and songs are petitions by the katsinas on behalf of all humans and the natural world to summon rain and its accompanying benefits.

A woman from San Ildefonso Pueblo gets water from the Rio Grande above White Rock Canyon in a 1905 posed photo by Edward Curtis. Edward S. Curtis, Courtesy Palace of the Governors Photo Archives (NMHM/DCA), neg. no. 144547.

About water, the late Hopi educator and linguist Emory Sekaquaptewa said,

> Water becomes the very basic element of life when it comes to Hopi religious prayers. *Everything* is focused on water and rain and rivers and underground springs—all of these things become part of the gift that the Hopis may use in order to sustain their way of life. . . . This has always been the position of the Hopi and always has been a principal belief focus of Hopi religion to this day.
>
> . . . For oral tradition [water's importance] has become a ritualized, formal acknowledgement of natural waters as a part of the *Tuuwapongya* . . . sand altar, which is the Earth. All things that are from the earth are a gift of our maker and as we live and discover their uses, it is for us to live on the earth and to exercise our religion in a sincere way so that these gifts will continue to come and see us as a people to our own destiny. That's what we promised [to Maasaw, the Hopi deity who is said to be the guardian of the present, fourth world], to try to uphold this acknowledgement of our maker and by living the Hopi way and praying with complete sincerity that we can continue this way of life forever, as long as we're allowed to live here on this earth.[50]

When the Hopis visit oceans and mountains, they bless them with sacred cornmeal, *hooma*, to bring rain. The oceans and mountains are acknowledged as the source of clouds and rain—a fact no modern meteorologist would dispute—and as such are revered as the wellspring of water, the foremost gift to the living world. Many Hopi songs mimic the sounds of diverse creatures— frogs, butterflies, birds—in their happy response to rain. Rather than being anthropocentric, the Puebloan universe is *hydro*centric, revolving around water, recognizing that all creatures and plants have an equal share in the living world only because of water. Mastery of water was never a goal.

Newcomers to the Land

The River When the Spanish Arrived

In 1598 Don Juan de Oñate marched out of Mexico at the head of a little army. On April 20, near what is now Ciudad Juaréz, he led the troops and colonists down to the banks of the Rio Grande. Across and up the river lay his field of conquest, the province of Nuevo Mexico that he would add to the Spanish Empire. But Oñate was brought up short, not by fierce resistance of the Indian population, but by the river itself. He gazed on a Rio Grande that will not be seen again in the contemporary world. The river was a foaming, seething torrent of water more than a quarter-mile wide. Two horses, their caution overcome by desperate thirst, plunged into the river to drink and were swiftly swept away to their deaths.[1] Oñate wandered slowly up the river for two weeks, looking for a spot that horsemen and wagons could cross, before the river finally subsided and he found a good ford at El Paso.[2]

The river that Oñate camped beside was a wide, shallow, braided stream that surged over its low banks nearly every spring when the snow in the San Juan Mountains melted. When spring floods were large, the river shifted its course, ripping up the vegetation along its banks and leaving its old bed an expanse of dry sand or a shallow wetland.[3] The bottomlands were covered by a patchwork of vegetation types. In many areas the river was linked to a network of swamps bordered by willows, cottonwoods, and aquatic grasses. In other drier but still-boggy areas, expanses of pale alkali grasses waved. Slightly higher and more stable parts of the landscape were covered by an open cottonwood and willow forest (called *bosque* by the invading Spanish), which loomed over a low understory of brush and grass.[4] The river and its associated wetlands were home to twenty-four species of fish, including sturgeon, gar, and eel that migrated up from the Atlantic Ocean to breed.[5] A large number of bird species nested or fed in the variety of habitats provided by the riparian vegetation mosaic. Wild turkey and whooping crane were among the more notable avian fauna. Beaver and mink inhabited the wetlands and numerous species of rodents lived in the drier habitats. These constituted the food supply for many predators, including even the occasional grizzly bear or jaguar.

Today at the same site and the same location, and during the same season, the Rio Grande is a muddy rivulet constrained between high levees. The native once-endemic aquatic species and riparian ecosystems that characterized this desert river are either now extinct or endangered. The difference between the Rio Grande we see today and the one that blocked Oñate's invasion is the result of centuries of reshaping the river to meet the needs of man. Oñate and his colonists were the entry point for the new ideas and new technologies at the start of that long evolution.

The first Spaniards to enter New Mexico in the mid-sixteenth century came to conquer the terrain and look for gold. This they did with a vengeance, as documented by the friars and soldiers that were among the first wave. But by the end of the sixteenth century, when the quest for gold was proving fruitless, the Spanish goals began to shift. Instead of just exploiting New Mexico's resources, the Spaniards settled the land. Instead of searching for mineral treasures, they looked for water. Instead of digging mines, they dug irrigation ditches and called them *acequias*. In terms of their relation to the river, the Spanish farmers who settled along the river—many of them soldiers turned settlers—and Rio Grande native peoples eventually found much in common, including gentle use of the river.[6]

The village of San Gabriel shows the pattern that was repeated time and again over the next four centuries. Spanish settlers picked a spot for their permanent homes at a place near reliably flowing water in north-central New Mexico, near present day Espanola. They located it at the confluence of the Chama River and the Rio Grande, across the water from what the Spaniards called San Juan Pueblo and what the residents call Okey Owingue in their Tewa tongue. Their first order of business was to make permanent life sustainable, and so they found a way to irrigate the land. Churches were important to their spiritual well-being and adobe houses were necessary for permanent shelter, but to them, the truly essential element for subsistence in the high southwestern desert was irrigation.[7]

Spanish American Irrigation: Acequias Come to the New World

In developing the essential irrigated agriculture, the early settlers at San Gabriel and at every Hispanic settlement established through the nineteenth century simply followed their cultural heritage. They came carrying the hydrologic baggage of Mexico, of southern Spain, of North Africa, and even ancient Rome. Their water vocabulary used words from the time of the Moorish occupation of Spain. They talked of *atarques* (diversion dams), *zansas* (irrigation duties), and *compuertas* (headgates) harkening back to the ancient Islamic world from which their irrigation institutions had sprung. When they began their settlements by constructing acequias, they drew on deep linguistic roots and their most fundamental concerns.[8]

Indeed, accessible water from rivers was the reason for the location of their villages in the seventeenth, eighteenth, and early nineteenth centuries. Moderate irrigation of marginal lands was indispensable to the survival of these villages. While the village acequias represented the first direct diversion of the Rio Grande basin's surface waters, they also carried a rich communal and religious meaning, which emphasized a healthy respect for the natural world and limited human power to control it.

Acequias provided the means by which the Spanish settlers could divert surface flows of New Mexico streams and apply the diverted water to agricultural lands so that crops could flourish. When laying out the land for their settlements, the first consideration was water and proving to the authorities there was an adequate supply.[9]

The lay of the land dictated where an acequia could go. In the upper Rio Grande basin, above La Bajada divide in the area just north of Cochiti Pueblo, the Rio Grande was so deeply entrenched in canyons that its water, except in the relatively small Espanola valley, couldn't be diverted and used for irrigation. Instead, Hispanic irrigation focused on the upper Rio Grande tributaries. There, the small valleys that had formed as the Rio Grande rift opened were arranged like beads along the string of mountain streams, relatively open places connected by the thread of narrow canyons through which the tributary streams rushed.[10]

These beads of valleys, with their open riparian land, proved the ideal setting for the simple irrigation technologies that the acequias embodied. Northern New Mexico tributaries also had manageable stream flows, especially compared to those of the much larger main-stem Rio Grande. At these elevations, the abundant natural precipitation supplemented irrigation flows during the growing season and offset a short frost-free growing season. At higher altitudes, the growing season would have been too short; any lower and the water supply would have been too powerful and erratic. Consequently, acequia culture took an early hold in the small valleys of northern New Mexico at elevations between 6,500 and 7,500 feet above sea level, because there the early demands of nature and agriculture could be most perfectly balanced.[11]

In 1590 the first Hispanic settlers had the forced assistance of fifteen hundred Pueblo Indians in digging the first acequias at San Gabriel. Thereafter, Spanish settlers usually dug the acequias themselves. They used primitive tools and limited technology to construct the ditches, but the objective was always the same: to bring as much land under irrigation as possible.

The construction strategy was similarly simple: community work gangs would go upstream as far as the terrain allowed in order to maximize the area that could be irrigated. There they built works that would divert the waters of a stream into the acequia.[12]

The acequia descended parallel to the river. Like the river, the acequia depended on gravity to carry its water down toward the open fields; thus, the system worked only if the drop in the line of the acequia was less than the drop of the river where the water originated. As the river and the acequia descended at different elevations, more and more vertical and horizontal space opened between them in order to maintain that drop ratio. By the time the acequia and the river emerged into a relatively open valley, the acequia would ideally be as far above and removed from the river as possible. The acequia continued along the upper rim of the valley while the river ran down through its center. Irrigated tracts lay in between. When the valley pinched closed downstream, the acequia rejoined the river.

Acequia del Molino

For such localized institutions, the acequias of the Rio Grande look remarkably similar. Walk the course of any one and you'll find the pattern common to them all. Take, for example, the Acequia del Molino in Cundiyo, New Mexico, which is north of Santa Fe, east of Nambe, and on the Rio Frijoles (see plate 11).[13]

The Rio Frijoles is a tributary of the Rio Santa Cruz, itself tributary to the Rio Grande. The town of Santa Cruz, near present-day Chimayo, was settled and its first irrigation ditches dug prior to 1695. The Acequia del Molino was probably dug before 1745, when the nearby Santo Domingo de Cundiyo grant was formally made. The first settlers there had applied to Spanish authorities for the land because the Rio Frijoles ran through the first relatively wide valley that the river had cut on its eighteen-mile course. It originated at almost 11,000 feet in elevation in alpine meadows and springs, ran down to a juncture with the Rio en Medio at 7,000 feet, and then joined with the Santa Cruz at 6,500 feet. Those first settlers set out to bring as much of the valley into agricultural production as they could by bringing water via the Cundiyo community acequia.[14]

Starting above the valley and on its east side, they must have first scrambled upstream across arroyos coming in from the east and past enormous boulders in an increasingly narrow canyon, home to a fast-moving mountain stream no wider than eight feet in most places. Half a mile upstream they found what they were looking for: a deep pool, formed by Rio Frijoles water cascading over boulders and coming to momentary rest beneath a huge boulder that loomed over the river on its east side. On the west side of the Rio Frijoles, the tight fit of the canyon continued, but below the pool on the east was a ten-foot-wide stretch of relatively flat land. There, in 1745, the builders of the Acequia del Molino saw pure opportunity.

They took advantage of this flat landscape by constructing the first of many loose-fitting rock dams across the Frijoles on the downstream side of that pool. The dam leaked, but it raised the water level just enough to force water into the ditch that they constructed just above the natural pool. From that starting point, the builders dug north along the east side of the north-running Frijoles, trying to stay as far from the river as the terrain and gravity would allow. The ditch measured no more than three feet wide and three feet deep. Lacking even the most rudimentary surveying equipment with which to check elevation, they would open the ditch from time to time to make sure that they were going downhill with a sufficient drop to produce the water flow needed.

Some difficult passages faced them. Perhaps two hundred feet below where they had dug their acequia, a massive boulder weighing tons lay right in their path. In 1745 the Hispanic work crew couldn't move it and the explosives to destroy it were well beyond their means, so they substituted hard work and skill for force. They laid the course of their new ditch right under the edge of that huge boulder, scarcely ten feet from the Rio Frijoles and five feet above it. In the margin between the two, a massive, tangled apple tree sprang up at some future time.

Another half mile down the acequia, the ditch evades a second physical obstacle. There the rocky hills on the east pinch in, leaving almost no terrain for Cundiyo's acequia to pass through. The builders of the acequia squeezed their ditch along a narrow ledge between the rocky hills and a sheer cliff overlooking the stream and made it just fit. More than 250 years later, the

Looking upstream on the Acequia del Molino near its diversion from the Rio Frijoles. The irrigation ditch runs under the huge boulder on the left. The river courses downstream just to the ditch's right. The budding apple tree hangs on in between. Photo by G. Emlen Hall.

western edge of the Acequia del Molino still runs the same narrow course, thirty feet above the faster-falling Rio Frijoles. The boulder, the ditch, and the fruit tree are all still there, testament to both the permanence and precariousness of the ditch.

Finally, the Acequia del Molino and the Rio Frijoles emerge together into the small, open valley that was always the target of both. The length of the ditch that conveys water to the valley is almost as long as the length of the ditch running on the east side of the irrigated fields.

There, Rio Frijoles courses north over a rock-and-pebble streambed as it runs through the center of the valley. The acequia, still parallel to the river but now ten yards above and two hundred yards east of it, hugs the highest valley contour that it can as it curls around the edge of the valley. By the time the walls of the canyon narrow again, a half-mile later, the Acequia del Molino and the Rio Frijoles frame almost ten acres of now irrigable land.[15]

For as long as anyone in the tight-knit, tiny Cundiyo community can remember, those ten irrigable acres have been divided into eleven tracts, running from the ditch to the river. None of the tracts are more than three acres in size, and some are less than one. The lay of this irrigated land has been the same for 250 years, supporting the community for a quarter of a millennium.[16]

Main Stem Versus Tributaries

Ditches just like Cundiyo's Acequia del Molino set the essential pattern for water resource development of the Rio Grande drainage, particularly in the upper Rio Grande in the Rio Arriba. There the best estimates show that between 1590 and 1846, Hispanic settlers built four hundred community irrigation ditches on upper Rio Grande tributaries that irrigated 55,000 acres, an average of about 130 acres per acequia, divided into tracts no larger than 10 acres.[17] Each system was unique to the water resource and terrain that gave it birth and to the community that built the acequia and modified the terrain. But all acequias were similar in many ways. All were located on tributaries of the upper Rio Grande drainage. They had no artificial storage in the form of dams. They depended instead on snowmelt at the beginning of the growing season and runoff from the more violent summer rains after those flows naturally dropped off. Acequias took water from the rivers, but because their technologies were not highly developed, their impacts were light.

Although these acequias were concentrated on the tributaries, the main stem of the Rio Grande offered too rich a resource to escape all upper Rio Grande irrigation development. The wide Espanola Valley offered particular promise. But even there only ten Hispanic acequias (less than 3 percent of all upper Rio Grande acequias) diverted water from the main stem, irrigating less than 3,000 acres historically and currently.[18] However, the fact that each main-stem acequia also irrigated almost twice as much land as the average tributary acequia in northern New Mexico suggests that main-stem irrigation allowed a broader scale of hydrologic development.

This suggestion is borne out by the history of irrigation in the middle Rio Grande area, from La Bajada (which separates the Rio Arriba from the Rio Abajo) to San Marcial. In that lower stretch of river, the tributaries are normally dry, rendering scarce the accessible water that made northern New Mexico so attractive. In the middle Rio Grande reach of Rio Abajo, the main stem offered the only reliable source of the surface water that was so indispensable to early settlement. Spanish settlement, and thus irrigation development, lagged in the Rio Abajo due to frequent attacks by nomadic Indians.[19] It was not until the early eighteenth century that irrigation began in this area, haltingly as it had in the north. The delay allowed techniques to be developed to divert the main river. This was rewarding, since here the irrigable flood plain was wider and the main stem volume much greater. There were also negatives: maintaining primitive brush-and-rock diversion dams against the force of the Rio Grande spring flood was challenging, while in late summer the river flows sometimes dropped so low that diversion was no longer possible. Nevertheless, the potentially irrigable land was much greater, and thus more expansive development was possible. No wonder, then, that there may have been as many as one hundred community irrigation ditches providing water to 140,000 acres by 1846.[20] Thus, the middle Rio Grande acequias on average delivered water to 1,400 acres, reflecting the basic differences between the Rio Arriba and Rio Abajo and their prospects for agricultural development.

Sharing the Waters

Acequias provided water for all uses. The classic residential pattern for these communities was to place houses uphill of the acequias. The acequias provided irrigation water to the fields below, while also providing relatively convenient domestic water to the houses just above.[21]

The acequias combined communal and private uses and obligations in ways that bound the communities together. Communal obligations included the care and maintenance of the diversions and ditches that served all the tracts under them. These communal obligations gave rise to apparently endless squabbles over the relative duties of individual members to the common ditch. But these squabbles also underscored the importance of the acequias and the necessity of joint work to maintain them. In spring, the ditches had to be cleaned and deepened. In summer, they had to be repaired from damage done by flood flows that ripped out the primitive diversion dams and by flash floods that destroyed ditches that traversed arroyos. After the autumn harvest, acequia members had to take down the temporary fences that divided the individual irrigated fields and open the seasonally fallow fields to communal grazing, following a practice originating in medieval Spain under the name *derotta de mieses* (grain rights) and carried to New Mexico under the more mundane *derecho de rastrojos* (right of stubble grazing). The annual round of communal work on the acequias never ceased.[22]

As for farming the irrigated tracts, the work fell primarily to the plot owner. Plowing, sowing, weeding, irrigating, and harvesting of crops: all were duties that belonged to the individual farmer in Hispanic New Mexico. Still, that work was intimately connected to the communal operation of the ditch, because the distribution of the water among the individual *parciantes* (members of the acequia with shared water rights) was governed by an acequia commission and a *mayordomo*, who basically served as a manager to carry out the commission's directions.[23]

Acequias rarely carried enough water so that tract owners served by it could irrigate whenever they wanted to or even simultaneously. This fact of acequia life wasn't simply a question of water quantity. It was also, and probably more important, a question of water pressure. The most efficient irrigation of a piece of land required the full flow of a ditch to push the water at right angles down the field. That was best done if irrigators took turns. To meet that fact of life, acequia governors had to control the distribution of water and its timing.[24]

No Spanish or Mexican law governed how each acequia should distribute water among its members. A remarkable range of methods and

A *parciante* (irrigator) diverts water through the *compuerta* (ditch opening) to send water to his fields in Rio Arriba County, New Mexico, ca. 1938. Courtesy Palace of the Governors Photo Archives (NMHM/DCA), neg. no. 58866.

considerations emerged over centuries of practice, especially in northern New Mexico. Sometimes ditch commissioners distributed water *por necesidad* (by necessity). When supplies were short and this principle was at work, ditch commissioners might favor garden plots, which provided daily food for families, over pasture. Or they might favor orchards, whose trees were permanent and hard to establish, over annual crops, which, if they failed in a year of drought, could be easily replanted the next.

At other times, acequia commissioners distributed irrigation water *por tandas* (in turns). Especially when river and acequia flows started to drop off in early June and before the summer rains came, this distribution required careful rotation of irrigation of individual fields in order to guarantee both an adequate supply for the field and an adequate pressure head for the irrigator. The larger the acequia and the more members, the more complex the rotation schedule became.

Finally, acequia commissioners sometimes distributed water *por derecho*, or by the relative rights of acequia members. On some acequias, irrigation water was distributed by the size of tracts under the ditch rather than membership, with the owner of the larger tract getting more irrigation time than the owner of a smaller one. Over the years and the centuries, some tracts came to have more rights than others because of size: the larger the tract, the bigger the water right. On some ditches, even more privatization took hold and rights came to have an abstract meaning, not even connected to land or size or membership. These rights were sometimes traded, even bought and sold, among members of an acequia, and commissioners and mayordomos distributed water accordingly.[25]

Prior to 1846 and for a long time thereafter, community ditches mixed and matched distribution by necessity, by turn, and by right in whatever way they saw most fit. A remarkable patchwork of practices emerged, a testament to the intricate and extensive regulation of essential acequia water once it left the rivers and entered the irrigation system. However, the rich practices of these institutions were so local and so hidden from the wider world that they were hardly noticed.

Within New Mexico agricultural communities, however, acequias were a central part of communal life, just as water had central importance in Pueblo life. In Hispanic New Mexico, the acequias were governed by a basic ritualistic Catholicism that emphasized the holiness of irrigation activities. As every Hispanic irrigator knew, San Isidro was the patron saint of farmers and his help was constantly needed and asked for. If help from God was not sufficient, then help from within the community would come in the form of the powerful mediating force of what anthropologists call a *shame* culture. To act against the water needs of the community was *sin vergüenza* (shameless), and there was no worse charge against a fellow parciante than that he *la dejo suelta*—let the precious irrigation water loose without exercising appropriate and required control over it.[26]

Isidro and Nepo: Saints of the Ditches

Northern New Mexicans have long paid homage to the centrality of water and irrigation in their lives and have woven it into their community ties through religious processions. Every community has its *acequia madre* (principal irrigation ditch) and its *capilla* (small chapel). Each of these has a mayordomo. Each capilla also has its saint, and the saint is usually San Isidro, the patron saint of farmers.

In the northern New Mexico version of San Isidro's rise to sainthood, Isidro was known as an upright individual who adored his Lord God. On one particular Sunday, Isidro was planting. God spoke to him, telling him that, as it was Sunday, he shouldn't work. Isidro responded that he had to work because it was the only day he could plant. The voice then warned him that if he didn't stop working, God would send him a ruinous hail. Isidro defiantly told the Lord that he wasn't afraid of hail and that he had to keep working.

In a booming voice, God then threatened Isidro again, this time telling him that if he didn't quit irrigating, He would send a plague of locusts to destroy his crops. Again, Isidro defied the voice and kept plowing and planting.

Then for the third and last time, God told Isidro that if he didn't quit working and go to mass, He would send him a bad neighbor who would be an uncooperative user of the acequia madre. Isidro could tolerate hail and locusts, but a bad neighbor was too much. He dropped his shovel, laid down his plow, and went to mass. When he returned, he was astonished to discover that an angel had planted and irrigated his fields.

For more than three hundred years, for each of the nine nights that precede the May 15 Feast Day of San Isidro, members of a community irrigation ditch take the statue of San Isidro from house to house. After Mass on May 15, the priest carries San Isidro on a procession through the community. The march proceeds along the just-planted fields and just-cleaned community irrigation ditches. The members of the modern community are his angels.

The next day, May 16, the farmers and parishioners celebrate the Feast Day of San Juan Nepomuceno, the patron saint of irrigation. Nepo, as he is affectionately called, gained sainthood the usual hard way, suffering martyrdom. He was imprisoned and tortured by King Wenceslaus IV for refusing to divulge the contents of his queen's confessions, then was drowned in the Moldau River, thereby earning the irrigation patronage.[1] In the wider Catholic world, Isidro is better known, but in northern New Mexico, it is Nepo who holds sway: after all, it is irrigation that allows farming in the desert. Hispanic parishioners repeat the local saying that *agua es la sangre de la tierra* (water is the earth's blood) as they open the lateral irrigation ditches (or *sangrías;* literally, the veins that come off the main blood vessel, the acequia madre) and spread water over the fields. With these two ancient ceremonies, the community each year celebrates both the labor contributed to maintain the acequias and fields and the divine grace that supplies the water, the lifeblood of the community.

Notes

1. Steele 2005, 377.

A Catholic priest and parishioners bless the water at the ditch that serves La Joya and Contreras, New Mexico. Photo by Nancy Hunter Warren, originally published in Warren 1987.

Hispanics and Pueblos: Two Approaches Grounded in Culture and Place

The Hispanic settlers of both the upper Rio Grande drainage and the middle Rio Grande shared their source of water, the tributaries and the main stem, with the Pueblos who were already in place when the Hispanics first arrived. Two hundred years from the first Spanish incursion into the area in 1540, Pueblo and Hispanic lands shared a common irrigated land use pattern. Each had an *acequia madre* (principal irrigation ditch) that diverted water from the Rio Grande or one of its tributaries. Each acequia served individual tracts of land. Some Pueblo ditches even served tracts of land that belonged to Hispanics who had either tacked themselves on to the end of Pueblo ditches or somehow got possession of a tract of land between pueblos on both sides.[27]

The shared acequia culture blurred the distinction between the Hispanic and Puebloan cultures, allowing them to overlap at least in this realm. Some Pueblos added a mayordomo position to help the village chief manage water and maintain ditches. At the same time, Hispanos modified the feudal style of water management that they inherited from Spain and adopted a more cooperative system of doing the communal acequia work, in line with their Puebloan neighbors.[28]

Despite the similarities, however, essential differences remained. Those differences showed primarily in how the ditches were internally operated and in the amount of control an individual had over his farmland. Public and private interests were never as evenly balanced in the Puebloan irrigation institutions as they were in Hispanic ones. Puebloan irrigation institutions were more infused with communal values. Despite some chafing, most pueblos recognized individually irrigated tracts as the property of the pueblo as a whole. Unlike their Hispanic neighbors, the individual families that worked a tract didn't fully own it. Pueblo officials could and did re-assign irrigated parcels among pueblo members. Assigned parcel users didn't keep for themselves all the harvest that their land produced. They were required to deliver a part of the parcel's yield to the community. And they couldn't freely sell or transfer assigned parcels without Pueblo agreement.

Within these few broad generalizations, for the most part the internal operations of all community ditches, both Hispanic and Puebloan, were detailed, intricate, idiosyncratic, and deeply determined by local culture.[29] The acequias hardly related to one another with respect to the waters of the Rio Grande as a whole, with few exceptions. Local governments occasionally struggled to define the relative rights of acequias, and some ferocious interacequia disputes broke out around the fast-growing Taos area in the early nineteenth century. But by and large the acequia commissions left one another to their own devices and customs.[30]

Roque Conjuebes: Santa Clara Rebel

From a distance, Pueblo Indian land use and irrigation resembled Hispanic institutions. Up close, there were real differences. They are illustrated by the long, bitter battle that emerged between the Santa Clara Pueblo and one of its own members, Roque Conjuebes, over irrigated land within the pueblo.[1]

By 1745 Santa Clara Pueblo had developed its own acequia system that physically resembled the systems of its Hispanic neighbors: a long community ditch, diverting water from the Rio Grande, ran south along the west side of the river and served tracts of irrigated land that were usually farmed by extended Pueblo families to whom Santa Clara officials assigned the parcels. But whereas Hispanic farmers owned the small tracts that they irrigated, individual Pueblo irrigators had a much less firm hold on the plots assigned to them, and the pueblo as a whole retained a much greater interest in the land.

This difference particularly vexed Santa Clara member Conjuebes. He maintained that the individual tract that he farmed was so tied up with communal obligations that he couldn't develop the land as he wished. In 1745 Conjuebes filed a petition of severance with Spanish authorities, calling these special obligations *vinculaciones,* a technical term of ancient Spanish land law referring to the medieval restrictions imposed by overlords on an individual's use of property. He portrayed the Santa Clara Pueblo leaders as if they played king to Conjuebes's serf—he had to give part of his produce to the pueblo, he had to meet the pueblo's ceremonial obligations—and he wanted to be free of them, a concept that contradicted the very foundation of puebloan life.

His petition asked Spanish authorities to essentially convert his parcel of irrigated Santa Clara land to non-pueblo land, which they promptly did. The officials freed his land from pueblo obligations and gave Conjuebes complete authority over it.

The decision never sat well with the pueblo and the debate over the status of irrigated land continued to simmer. In 1813, sixty years later, Santa Clara Pueblo applied to the high court in Guadalajara, Mexico, five hundred miles away, to reconsider the 1745 decision. That court reversed the decision and ordered the Conjuebes family and descendants either to rejoin Santa Clara Pueblo, with its attendant obligations, or abandon the land and move elsewhere.

The 1813 affirmation of the communal rights of the Santa Clara Pueblo wouldn't be the last time that a court would consider the balance between the joint rights of the pueblos and the individual rights of Pueblo and Hispanic farmers of the Rio Grande. But the case of Conjuebes illustrates a fundamental difference between Pueblo and Hispanic philosophy about natural resources. Hispanic land in New Mexico is privately owned and farmed, with water carefully shared through a delicate and elaborate system of community cooperation. To the Pueblo people, land is primarily a communal resource that may be tended for decades by an individual, but only as a steward. The land ultimately belongs to the people and the clan, and water is a gift to all.

Note
1. Undated 1744 petition of Roque Conjuebes to Governor Codallos y Rabal ending with August 1816 decree (Twitchell 1914/2008, vol. 1); see Hall (1988).

Water Supply and Water Scarcity

Despite the rich history and careful development of these internal legal principles, acequias did not much figure in the larger political arena until the late nineteenth century. The fact that New Mexican community acequias were so ubiquitous and deeply determined by local culture and concerns may explain this lack of a specific legal identity before 1895. It was hard for acequia groups to find a supervising, regional legal authority. A fuller explanation for their obscurity in the political arena, however, may be the sheer physical reality they faced: each ditch was much more at the mercy of the river that fed it than it was at the mercy of other ditches.

In the Rio Arriba, tributary streams were diverted by means of primitive dams. Storing water upstream or capturing the full rushing mountain flows was rarely possible because of the terrain. Downstream irrigators on the same tributary received the water that escaped the crude upstream diversion along with the surface water and groundwater that flowed into the stream below the first diversion. When a ditch went without water, it was much more likely due to the fact that there was no water in the river rather than that another ditch had taken more than its fair share.

In the middle Rio Grande, the facts of water life were a little different. The volume of water there was much greater than in the northern New Mexico tributaries. But the variation in flow was much greater too, and the phenomena of water feast and water famine were more pronounced. Huge runoff flows from melting snows at the beginning of the irrigation season wreaked havoc with Rio Abajo diversions, often requiring them to be rebuilt several times within an irrigation season or even abandoned. The task of rebuilding diversion structures across the swollen, snowmelt-fed river was both dangerous and numbingly painful. Further, the constantly shifting channel of the middle Rio Grande offered a moving target that produced a rich and unique biota but a poor supply of irrigation water. And if that were not enough, the middle Rio Grande often simply dried up toward the end of the summer due to waning snowmelt, not excessive diversion.

In the absence of technology to deal with these problems, pre-1846 farmers resigned themselves to a water supply on nature's terms. This was a world in which the rivers—the works of nature—had primacy. The ditches were the works of men, and as rich as the communities that created them were, those ditches came second. This was simply the way it was.

New Mexicans in the Rio Grande drainage expressed this sentiment in one of the time-honored *adivinanzas* (riddles), which, generation after generation, defined the Hispanic world. *¿En qué es suspendido el mundo?* the riddle would begin, asking how the world was hung in the universe. The answer: *En la voluntad de Dios* (by God's will). *¿En qué es suspendido nuestro pueblo?* (And how is our pueblo supported?) the riddler would continue. *En el río* (by the river) was the standard answer. In nineteenth-century New Mexico, God, and the Rio Grande were equally central, powerful, autocratic, and unmanageable by humans. The church managed God's affairs on this earth; the acequias managed whatever water they could capture. But make no mistake: God did his will. The river did its will too, pretty much unmanaged by man.[31]

Until the 1940s all the essential values of the northern New Mexico Hispanic communities coalesced around their acequias. The acequias were the focus of their religious life, the central activity of their communal life, and absolutely crucial to their physical survival in incredibly isolated high desert towns. Since then, some of that centrality has been lost. The small tracts of irrigated land are often not large enough to be economically productive. Wage labor is often required for the maintenance of a family, and farming becomes a weekend or retirement activity. But a powerful religious and spiritual attachment to the acequias and the land remains, despite a changed Rio Grande basin.

Even now, stickers reading *el agua bendita* (blessed water) are plastered on the bumpers of pickup trucks traversing the roads of northern New Mexico villages. The stickers testify to the enduring importance of water in these Hispanic communities.

Chapter IV

A New U.S. Regime

In August 1847 General Stephen Watts Kearny and his Army of the West marched into the plaza at Santa Fe, New Mexico, and without a battle officially took possession of the territory embracing the Rio Grande basin. Kearny's triumph was one of the easiest victories in the Mexican-American War through which the United States conquered the territory of New Mexico. The coming of the United States, however, was also the beginning of a different kind of conquest. The rivers that Hispanos and Pueblos considered beyond their control would eventually be harnessed by the new society, using new technologies and new laws.

This type of conquest was ushered in by the wave of Anglos who would follow Kearny's path westward and was based on a particularly American philosophy about land and natural resources. The importance, even responsibility, of the individual to achieve success and realize profit overrode obligations made by the new government to the existing communities. Subsistence economics was scorned by this new wave of immigrants, who sought fortune through the timber, cattle, and railroad industries in the name of fulfilling the destiny of a nation. The consequences of these enterprises would soon begin to collect in the Rio Grande basin, trickling down into the river and accumulating quite literally in a heavy load of eroded soil, sediment, and silt.

The Mexican-American War won a huge territory for the United States, but also resulted in a treaty obligation to honor the existing land and water rights. Kearny quickly moved west, across the Rio Grande to the more coveted, although less populated, California. In his wake, Kearny left a new national sovereign, a small force, and a profound ignorance about the irrigation-based culture of the Rio Grande basin.

Before he left, Kearny adopted a military code of dubious legality, which simply provided that "laws concerning water courses . . . shall continue in force."[1] A regularly constituted territorial government established in 1851 and dominated by native New Mexicans confirmed this principle in the 1851–1852 sessions of the New Mexico Territorial Legislature, validating acequia practices without defining or regulating them. These early acts established a pattern that would continue through to the twentieth century and beyond: whenever the new government ruled on or made changes to the Rio Grande, it left the internal governance of the acequias alone. Acequias weren't worth the time or attention.

The new federal government was focused on the future; traditional subsistence and cultural practices were not its concern. Agriculture in general was not particularly of interest at first. Even before the U.S. seizure of the land, traders and adventurers were the principal authors who wrote about New Mexico in the relatively few and occasional reports that were published. In 1851–1852 the new U.S. government managed to drum up enough enthusiasm to formally survey existing agricultural lands in a series of reports that detailed crop production.[2] These publications provided more statistical detail than the sketchy earlier observations, but none of them showed much awareness or understanding of the local culture centered around the acequias.

Manifest Destiny: A Change in Perspective

During the first two decades of U.S. control of New Mexico, changes in Hispanic and Puebloan societies were relatively small. However, the uneasy yoking of these more traditional cultures with the very different agenda of the newest conquering state set the stage for conflicts that would play out over the next century. The Hispanic culture of New Mexico had a fundamentally local focus. The village was the root social unit, and collective decisions were based on how they would affect the village. This was true to an even greater extent for the Native American cultures. In contrast, Anglo American culture tended to be individualistic and directly linked to grand, national-level social movements. In 1839 John L. O'Sullivan wrote in a newspaper essay:

> The expansive future is our arena, and for our history. We are entering on its untrodden space, with the truths of God in our minds, beneficent objects in our hearts, and with a clear conscience unsullied by the past. We are the nation of human progress, and who will, what can, set limits to our onward march? . . . For this blessed mission to the nations of the world, which are shut out from the life-giving light of truth, has America been chosen; and her high example shall smite unto death the tyranny of kings, hierarchs, and oligarchs, and carry the glad tidings of peace

and good will where myriads now endure an existence scarcely more enviable than that of beasts of the field. Who, then, can doubt that our country is destined to be the great nation of futurity?[3]

The essay was entitled "Manifest Destiny," and the almost messianic social fervor it conveys was a major factor in justifying the military annexation of New Mexico a decade later.

In the minds of the new American colonizers, the divinely ordained destiny of the nation went far beyond territorial conquest. America was commissioned to establish a paradise on earth. This paradise would be founded on the fruitful exploitation of natural resources that had been underutilized by former inhabitants due to ignorance or indifference. One shining opportunity was the vast expanse of untilled land in the western portion of the continent. A bountiful supply of food was to be produced on what had previously been considered wasteland. The application of advanced American agricultural practices would be an important means to fulfill the destiny of the nation.

This agricultural destiny was foreseen to be achievable in an almost miraculous fashion. Much of the western region was clearly too arid to produce reliable crops. This limitation was widely expected to be overcome simply by ignoring the climate and farming anyway. The force of Manifest Destiny was so powerful that it could be counted on even to modify the climate. The theory that "rain follows the plow"[4] was legitimized by early climatologists (most notably Cyrus Thomas); it postulated that the act of plowing the land would, through rather mysterious means, cause increased rain to fall. No less an authority than Ferdinand V. Hayden, the head U.S. geologist, asserted in 1869: "I therefore give it as my firm conviction that this increase [of rainfall] is . . . in some way connected with the settlement of the country; and that, as the population increases, the amount of moisture will increase."[5] It was taken up and popularized by traveling diarists such as Josiah Gregg:

> Why may we not suppose that the genial influences of civilization—that extensive cultivation of the earth—might contribute to the multiplication of showers, as it certainly does of fountains? Or that the shady groves, as they advance upon the prairies, may have some effect upon the seasons? At least, many old settlers maintain that the droughts are becoming less oppressive in the West. The people of New Mexico also assure us that the rains have much increased of latter years, a phenomenon which the vulgar superstitiously attribute to the arrival of the Missouri traders.[6]

The idea that an individual could play a role in a national epic destiny, combined with the pseudoscientific assurance of a future with abundant rain, caused tens of thousands of Americans to migrate from the well-watered eastern United States onto the dry rangelands of the West. But when wishful thinking proved unsuccessful in modifying the climate, these new settlers ended up broke and busted.

By the end of the nineteenth century, the futility of commercial dryland farming in a semiarid region was apparent to most. But this realization didn't quash enthusiasm for agricultural millennialism; it merely rechanneled the fervor into "making the desert bloom."[7] The new frontier remained undeniably attractive. New Mexico's population had more than doubled between 1850 and 1890.[8] By 1910 it increased fivefold and 40 percent of its 327,301 citizens had been born in other places. The destiny the new settlers envisioned would not be accomplished without remaking the Rio Grande into a very different river than the one it had been for thousands of years. Destiny would be fulfilled by diverting the unused waters of western rivers onto fertile lands where an agricultural paradise was sure to spring up. Unlike the idea of rain following the plow, which required only optimism and the almost inexhaustible resource of individual American initiative, this approach required extensive investment in technology, capital, and social organization.[9]

Until 1849 New Mexican society had been both locally focused and very technologically conservative, attitudes that had proved successful in facing an environment only marginally suitable for agriculture. This traditional perspective was now confronted with an alien society that had an unshakable conviction in divine destiny and a willingness to take whatever risks and make whatever changes were necessary to ensure that destiny would be realized. These newest immigrants brought new ideas about everything, including a tendency to think of Hispanic New Mexicans as "backward." They arrived with their heads full of ambition, myth, and dreams of creating a midwestern paradise in the Rio Grande basin. They would make the desert bloom, and booming economic development would inevitably follow.

The Railroads Arrive

On the frigid winter evening of February 26, 1878, in Raton Pass, New Mexico, Richens ("Uncle Dick") Wootton was annoyed to hear someone calling his name as he headed off to bed. Wootton had constructed the first toll road over the pass and was the owner of a rather primitive hostelry that served the road. When he turned, he was surprised to see William R. Morley and A. A. Robinson, the two top engineers for the Atchison, Topeka, and Santa Fe Railroad (AT&SF).[10] Wootton must have been even more amazed and perhaps hesitant when they told him that they would pay him to assemble a work crew to begin constructing a railroad grade over the pass that very night. When they next told him how much they would pay for a night's work, his hesitancy disappeared. He immediately started pounding on doors and rousting the sleeping residents. By two o'clock in the morning, he had an amateur crew assembled, and in the darkness they started up the mountain-side toward the pass. It was four o'clock before Morley said, "This'll do."[11] They formed a line and shoveled away at the thin alpine soil.

Raton Pass, crossing the easternmost spur of the Rocky Mountains, held the key to penetrating the Rio Grande valley in New Mexico. As the impromptu laborers (mainly Anglo Teamsters and the laborers who ran the hotel, stage stop, and stables) began digging up the pass, the rest of the population of New Mexico slumbered under the moonlight. This population consisted mainly of Hispanic subsistence farmers living much the same life that their ancestors had for hundreds of years. They felt none of the competitive urgency that drove the engineers to hack away at the mountainside through a freezing winter night, but they would eventually suffer the effects. Within a human life span, both the pristine Rio Grande and the ancient lifeways of

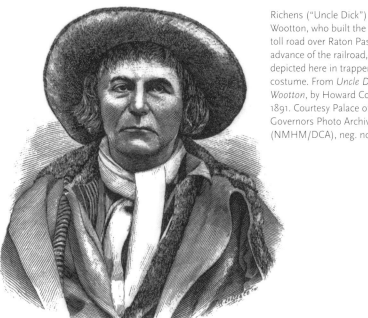

Richens ("Uncle Dick") Wootton, who built the first toll road over Raton Pass in advance of the railroad, is depicted here in trapper's costume. From *Uncle Dick Wootton*, by Howard Conard, 1891. Courtesy Palace of the Governors Photo Archives (NMHM/DCA), neg. no. 133216.

"UNCLE DICK" IN TRAPPER'S COSTUME,

those who lived on its banks would become irrelevant, swept aside as a wave of change flowed over Raton Pass.

What did the strange midnight work party mean? It was a question of title. After shoveling for only a half hour, Morley and Robinson saw lanterns approaching from the darkness below. In their light, a work crew became visible carrying picks and shovels. "Must be the Rio Grande boys!" Morley gleefully announced.[12] He was delighted to have succeeded in beating them to the punch on behalf of his own company. No one had yet formally claimed possession of Raton Pass, although the Denver and Rio Grande Railroad (D&RG) had surveyed a route two years earlier. The previous evening Morley and Robinson had spotted the D&RG men coming into town and bedding down for the night. In order to lay their claim to the pass, Morley and Robinson had immediately rousted Wootton and raised a crew on the spot. By means of their midnight excavation, they had staked first claim to the railroad route over Raton Pass and, from there onward, into New Mexico and Texas. A night's labor had earned them first rights to an inland empire.

The D&RG engineers were not defeated for long. To them, the loss of the only tractable pass between Denver and the Rio Grande was merely a momentary setback. The D&RG swiftly changed paths and drove its rails south to the Rio Grande through the San Luis basin, reaching Alamosa, Colorado, in June 1878. The engineers did not pause, but excavated elaborate grades over the high Conejos Mountains to reach Durango, Colorado, then continued southward to return to the Rio Grande in the Espanola Valley at the end of 1880.[13] This high and lonely narrow-gauge railroad through the Hispanic outposts of southern Colorado and northern New Mexico would eventually come to be known as the Chili Line. At the same time a third railroad, the Southern Pacific (SP), was building eastward from California. It reached the western border of New Mexico near Lordsburg in September 1880. By May 1881, the SP had extended its tracks all the way across southern New Mexico and the Rio Grande to El Paso, Texas. Meanwhile, the AT&SF continued to charge southward from Raton Pass, reaching Albuquerque in April 1880 and finally linking up with the SP at Deming, New Mexico, on March 8, 1881.[14] Thus, in the space of almost exactly three years, New Mexico transport leapt from oxcart to railroad. The Rio Grande valley was suddenly crisscrossed by rail, the most advanced form of transportation of the day (see plate 12).

The passion that drove this frenzy of effort was both personal and public. The personal drive ranged from the anticipation of venture capitalists that their risky railroad investments would return untold wealth and secure their own piece of the national destiny, to glad acceptance of backbreaking work in the wilderness by uneducated men, in return for fat monthly paychecks.

But the spectacular success of the railroad expansion was driven as much by public enthusiasm as personal benefit. Both the Spanish incursion north into New Mexico and this new invasion of Anglo Americans were driven by desire for wealth and treasure, but the Spanish aimed also to spread Catholicism throughout the New World. The corresponding Anglo passion was for the dissemination of new technologies and scientific progress. To the entrepreneurial Anglo Americans who had recently moved into the towns and villages of the Rio Grande valley, the railroad represented their access to markets, their engine of growth for jobs and industry, and their link to civilization. The railroad barons saw themselves as conferring the benefits of the latest advances of technology upon the population, a contribution that would better people's lives in every way. No one, it seems, found reason to voice objection to this grand vision.

The expectation of great change was certainly fulfilled. Some of the change was as anticipated, in the form of much faster and more comfortable personal travel, new markets for local products, and growth in industry and employment. However, the sudden appearance of the railroad also changed the lives of the inhabitants, the landscape, and the river in ways that no one had foreseen. Many of these changes were detrimental and irreversible, and we are still attempting to deal with them to this day.

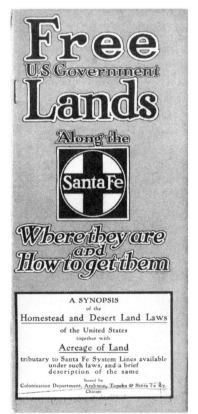

One immediate consequence was a radical, virtually perpendicular shift in the geographic poles of New Mexico's social and institutional orientation. Since 1598 that orientation had been north-south, with institutional authority, social attitudes and trends, and manufactured goods flowing north from Chihuahua, Mexico, to Santa Fe, and agricultural goods and provincial citizens in search of the wider world moving southward. This axis started shifting in the 1830s with the

Brochure published by the Colonization Department of the Atchison, Topeka, and Santa Fe Railroad in June 1906. Atchison, Topeka & Santa Fe, Printed Material file 2, courtesy of University of Arizona Libraries, Special Collections.

opening of the Santa Fe Trail, but completely rotated east-west after 1880, as authority came to be transmitted from Washington, D.C. Vast amounts of goods moved between St. Louis, Missouri; Santa Fe; and California, and American culture swept in from the east and west coasts with lag times of a few days, compared to the months and years that transmission took in pre-railroad days. New Mexico and the Rio Grande were now wide open to the imperatives of a national destiny.

Fate of the Forests

Some of the impacts of the railroad were direct and immediate, such as to the forests of New Mexico. At the time the railroads were built, these forests carpeted the mountainous headwaters of the Rio Grande, where most of the flow in the river originated. Forests regulated the inflow of water and sediment to the river, although at that time almost no one appreciated the implications of that fact. Before 1878 wood had mainly been used in small amounts for fuel, for roof beams (most houses in New Mexico were constructed from adobe), and for fences and implements.

Chaco Deforestation

While the extent of timber harvesting in New Mexico before the development of the railroads was generally quite modest, a notable local exception was the large deforested area around the ancient settlements of Chaco Canyon. Studies of packrat middens at Chaco show that prior to the establishment of the first Chaco Anasazi settlements a little before AD 1000, piñon pine and juniper were well established in the area. The first Chaco settlers soon began to harvest the piñon and juniper to use in construction and as firewood for cooking and keeping warm. Very quickly, by AD 1000, the area was stripped bare of the piñon and juniper woodland by the burgeoning, concentrated population.

Great-house timber demands and fuel requirements forced the Chacoans to find more remote sources for their construction needs, and so the deforestation spread outward a great distance. Ponderosa pine, spruce, and fir logs that weighed up to seven hundred pounds were subsequently chopped down with stone axes and carried twenty to fifty miles to be used as roof beams for the great houses. Chacoan loggers hauled ponderosa from the Chuska Mountains to the west, the La Plata and San Juan mountains to the north, and possibly Hosta Butte to the south, and took spruce and fir from the Chuskas and to a lesser degree, from the San Mateo Mountains to the south.

More than two hundred thousand trees were felled and transported to fill the needs of the Chacoan cultural complex until it collapsed and construction abruptly halted. Deforestation and water management impacts are thought to be the primary environmental challenges that the Anasazi brought upon themselves.[1]

Note
1. Midden studies by Betancourt and Van Devender (1981); isotopic analysis of timber sources by English et al. (2001), as cited in Diamond (2005, 147–48) and A. Reynolds et al. (2005). See also Stuart (2000, 78) and Solstice Project (1999) for estimates of the numbers of logs required in great-house construction. An alternate view is presented in M. Wilcox (2010, 128–33).

The new railroad had a voracious appetite for wood. For example, when a second set of tracks was constructed by the AT&SF across northern New Mexico in 1914, the project required more than 16 million ties.[15] The bridges and trestles of the day were elaborate laceworks of wooden beams, each one requiring enough trees to build dozens of modern houses. Once the railroad was built, the train engines devoured wood at a tremendous rate in their boiler fires. Industries enabled by the railroad consumed additional wood. For example, the railroad opened up New Mexico to industrial-scale mining, which required large amounts of wood for timbering to keep the mine shafts and tunnels from collapsing. Smelters required even more wood. As the ripple effects of industrialization and population growth expanded, wood consumption grew to a peak of 675,000 cords in 1918.[16] The cutting of such vast amounts of timber profoundly affected the landscape.

Where did all this timber come from? One answer to this question illustrates the ongoing collision of civilizations. A large part of the forest resources of New Mexico were locked up in Spanish and Mexican community land grants. The large expanses of community land played a critical role in the survival of the small Hispanic communities of northern New Mexico. In the summer, everyone was entitled to drive their livestock up to the mountain pastures to graze. Anyone could cut the trees for winter firewood or to build houses or fences. Without this resource, the members of the community would have been unable to survive. Under the Treaty of Guadalupe Hidalgo, the U.S. government was bound to honor these grants.

The newcomers had a wholly different viewpoint, as explained by writer and conservationist William duBuys:

> To the Hispanics of New Mexico the land was not a sacred thing as it was for the Pueblos, but it was still not altogether inanimate. It was the

A logging crew working near Santa Fe for the Atchison, Topeka, and Santa Fe Railroad poses for posterity. Photo from W. M. Beiger Collection, Archives 116, Small Collections Box 2, Center for Southwest Research, University Libraries, University of New Mexico.

mother and protector of their traditional subsistence pastoralism. In many cases it was a communal thing, belonging not just to individuals but to whole villages as a collective possession.

Initially, in the Anglo view the land possessed neither spirit nor communal qualities. Land [and trees] was simply a commodity, which like wheat or iron ore might be advantageously bought and still more advantageously sold.[17]

However, before the land could be "advantageously bought," a problem had to be solved. What land was under private title before the United States assumed control of the territory and so was not part of the new public domain? Could a former sovereign's concept of property be retained under U.S. law? Could land titles that had originated under Spain and New Mexico and that had been guaranteed under the Treaty of Guadalupe Hidalgo by the most recent sovereign, the United States, be sufficiently defined to make purchases possible?

The answer is yes, but not wisely or well, and not in a principled or even ethical manner. After 1848, the United States made a mess of Spanish and Mexican land grants, particularly those granted to communities.[18] The United States wrongly rejected some grants. It cut others down in size. It

The "S" bridge constructed near Cloudcroft, New Mexico, shows the massive amounts of lumber required in early railroad construction. Photo by Jim Alexander, courtesy Palace of the Governors Photo Archives (NMHM/DCA), neg. no. 11885.

confirmed ownership of still others to the wrong individuals or entities. The process created a legacy of suspicion and bitterness that continues to this day. But the process also created titles to land grants that were sufficiently clear to permit the American newcomers access to the resources on the lands, especially timber.

A classic example of injustice commonly occurred when the United States confirmed ownership of the community land grants not to the community but to the heirs of the original grantees. Land that had been granted to named community representatives in the late seventeenth century might have thousands of heirs by the time an ownership battle reached the courts. Under New Mexico law in the late 1880s, any one of those heirs had the right to sever his individual share. A member of the community would be financed by outside interests to sue the rest of the community for his share of the property. A court friendly to the outsiders would find that it was not possible to physically separate one individual's share, and therefore the entire grant would have to be sold and the proceeds distributed to the owners.[19] Outside interests would obligingly step up to the plate and offer a very modest amount for the grant. The fact that nearly all inhabitants of the lands in question opposed its sale made no difference. The end result of this legal maneuvering was that by the 1940s, up to 80 percent of lands thought by Hispanic residents to belong to their communities had been lost to outsiders.[20]

The Santa Barbara land grant in northern New Mexico offers one such example.[21] After the United States confirmed the grant, a suit for partition was approved, and the grant was purchased in 1907 by a New Englander named Amasa B. McGaffey. He set up the Santa Barbara Tie and Pole Company, which began in 1909 to systematically cut the entire forested portion of the grant. The trees were floated down the Rio Grande to Cochiti, where they were hauled to a creosoting plant in Albuquerque and turned into ties that were sold to the railroad. McGaffey abandoned the enterprise in 1926 after he had stripped the entire grant bare.

The effect of such logging practices was devastating. The mixed conifer forests of the highlands had very effectively slowed down the runoff and

Wooden railroad ties jam the Rio Grande in 1915. Courtesy Palace of the Governors Photo Archives (NMHM/DCA), neg. no. 039350.

held the soil in place. With the trees stripped away, the runoff became flashy and intense. Running over denuded land, these intense flows eroded the soil and flushed it into the Rio Grande and its tributaries. Floods became more frequent and intense, and between the floods, the bed of the river became choked with the silt that had been flushed from the deforested mountainsides. The inhabitants of lowland New Mexico would soon begin to pay a heavy price for the activities that were so profitable for McGaffey and his associates.

A Plague of Cattle

The cattle industry was also made great in pursuit of Manifest Destiny, and it, too, destroyed the relationship between land and water that had sustained the old ways. While the highlands of the northern Rio Grande basin were covered by dense mixed-conifer forest, the large expanses at mid-elevations supported open ponderosa pine forest. The soil between towering, widely spaced pines was anchored by a thick cover of grass. The ponderosa pines were certainly heavily logged, but one might think that, in this environment, the grass would protect the land and the ponderosa would recover. Such was not the case. Today, large expanses of the northern Rio Grande basin that formerly were covered by ponderosa savannah are under dense juniper and piñon woodland. Beneath these small, closely spaced trees is little grass, and the soil is eroding rapidly.[22] Although the cause can ultimately be traced to the railroad, that was not the primary culprit.

In 1870 the grazing lands of the upper Rio Grande watershed in New Mexico supported about 30,000 cattle and 350,000 sheep and goats.[23] By 1900 these numbers exploded to 150,000 cattle and 1.6 million sheep and goats.[24] How did this happen? Previously, livestock primarily supplied the meat, milk, and wool needs of just the local New Mexican inhabitants, although a small portion of the animals were driven to distant markets by Texas cattlemen. Suddenly, in 1880, those distant markets could be accessed by merely loading the animals into cattle cars and shipping them wherever prices were highest.

Technology allowed greed to flourish. The land had already begun eroding from centuries of tree cutting and overgrazing of sheep by Hispanics and Pueblos; the commercialization of the livestock industry, combined with the deforestation caused by the coming of the railroads made it dramatically worse.[25] Communal grazing land that was of interest only to nearby villagers now became an avenue to a fortune, if only it could be accessed. And accessed it was, by suits of partition and similar means. To maximize profits, the new livestock barons crammed onto their land as many animals as they possibly could. The temptation of grabbing short-term gains at the expense of long-term rewards was apparently too compelling for those who had both the opportunity and means to take advantage of the situation.

The livestock numbers were, of course, far beyond the sustainable carrying capacity of the range, but the damage actually started well before the grass was gone. Grazing was largely responsible for the change of landscape from park-like ponderosa savannah to dense stands of piñon-juniper. Before heavy grazing, summer fires sparked by lightning would burn quickly through the grass, eliminating most of the pine seedlings but sparing the towering, thick-barked, mature ponderosas. With the grass thinned by overgrazing, these fires no longer flamed across the landscape, and innumerable juniper and piñon seedlings—which are not eaten by livestock—took root. The fertile grasslands were quickly replaced by dense stands of small trees with bare soil beneath their thick canopies.[26]

The forested areas that had been denuded by logging and the overgrazed ponderosa zones now reacted differently to precipitation: erosion immediately and dramatically increased.[27] Rain had less chance to soak into the ground, and storms were more likely to produce the intense runoff that eroded the soil. In a few decades, the fertile soils that had taken thousands of years to develop were eroded off and dumped into the streams and rivers at the headwaters of the Rio Grande. The changes in the pattern of runoff and the amount of sediment in the system caused the river itself to begin to behave differently than it had for the past centuries, and these changes posed a threat to the populations downstream.

Chapter V

The River Pushes Back

Until the coming of the railroad, although a substantial number of people lived in the Rio Grande basin, their presence was felt relatively lightly on the land. Once the railroad allowed industrial society to rush into the technological vacuum of the region, conditions started to change. The river that had previously offered a reliable supply of irrigation water began to go dry for months at a time. Towns that had been well above the level of the river began to be battered by flood after flood. More insidiously, fields began to slowly sink beneath pools of stagnant, salty water. What was happening? Were these disasters somehow linked? Although the inhabitants of the Rio Grande valley did not initially make the connections, the river was beginning to respond to the demands of the new industrial society.

The Mystery of the Missing Water

As the end of August 1896 approached, the small farmers in the little villages west of Albuquerque often did not even bother to head out to their fields in the morning. With no water in the ditches, there was no irrigating to do, and looking at the fields was too depressing. The flow in the Rio Grande had shrunk to a trickle by the middle of June, just as the seedlings were getting established, then dried up altogether. The only reason anything was alive in the fields was the occasional downpour from a passing thunderstorm, but that amount of rainfall was nowhere near sufficient to sustain most crops. The vulnerable chile peppers and squash were withered and gone, and just a few stunted corn plants hung on under the scorching summer sun. The only crop likely to actually yield a harvest was the hardy bean plants.

A cheerless autumn and a hungry winter lay ahead. What had gone wrong? For as long as anyone could remember, the Rio Grande flowed reliably past the villages. In very dry years, the flow might vanish for a few weeks in July, but it would soon come back. Now the river was often dry for months each summer. The transition from good times to bad had been sudden. It had happened seven summers previously, in 1889, a year when the river had dried in late June.[1] The river had produced a good flow during a few summers since then, but they were the exceptions.

Was God angry? Were they being punished for abandoning the old ways? Where had the water gone? The answer they would eventually realize was that the railroad had come through the distant San Luis Valley, two hundred miles to the north. With it had come changes that would begin to strangle the Rio Grande, upon which so many New Mexicans depended for their livelihood and their food.

Hispanic settlers had first moved into the central San Luis Valley of south-central Colorado in the 1850s and began digging irrigation diversions almost as soon as they arrived.[2] As had the settlers farther south, they stayed away from the Rio Grande itself. Their flimsy brush diversion dams could not withstand the roaring torrent of spring snowmelt, and working in the fast-flowing and icy river to repair the diversions was torturous and life threatening. Instead, they stuck to the minor tributaries where small diversions were practical to build and maintain.

The early Hispanic settlers in the San Luis Valley lived a harsh life, far from the nearest civilization and constantly under threat of death or enslavement by raiding Ute Indians. They had struggled to survive the arctic winters that grip the basin. This situation changed in a matter of months in 1878, when the Denver and Rio Grande Railroad charged down the length of the valley. All manner of foods and goods could now be imported from Denver—or even the remote Atlantic coast—in a few days to a few weeks. Unfortunately, the flood of goods was inaccessible to the Hispanic settlers, who had very little money to buy imported products.[3] Such was not the case for the capital-rich Anglos, who stepped off the train looking for opportunities. In the space of a few years, they built sawmills, quarries, freighting firms, mining companies, and construction businesses where formerly there had only been wilderness.[4] Any needed supplies could be obtained by railroad.

Attack of the Monster Canals

Within a year or two, the most ambitious of the new empire builders realized that a new opportunity for substantial wealth lay before them. As the Rio Grande emerges from the San Juan range at the town of Del Norte, Colorado, it begins to flow southeast at an angle to the front of the range, and thus the topography rises fairly rapidly to the west. On the east, however, the river is bounded only by a low, almost imperceptible ridge a few miles into the San Luis Valley. Beyond this ridge the land slopes gradually downward to the east until it terminates in a swampy depression against the Sangre de Cristo Mountains (see plate 2). It is, in fact, a closed depression, the bed of the ancient lake that occupied the valley up until about four hundred thousand years ago.[5] The entrepreneurs saw that by contouring canals over the gentle ridge they could direct the water down the west flank of the valley and irrigate vast acreages that had previously been used only for grazing livestock.

Out-of-state sources of capital, most notably the Travelers Insurance Company headquartered in Minnesota, quietly bought up the land in the valley for low prices and began to construct permanent diversions and dig deep canals.[6] Diversion works that had been impossible for the Hispanic settlers were built in a matter of months with the aid of construction crews numbering in the hundreds, steam shovels, poured concrete, and steel. Construction of the Rio Grande Canal began in 1882, but it was 1886 before substantial amounts of water were diverted. Within five years, this enterprise was followed by the Monte Vista Canal, the Empire Canal (the builders were not shy about advertising their ambitions), and the Farmers' Union Canal. All of these diversions came online between 1886 and 1889.[7] Once the canals were underway, the developers began to sell plots of farmland for prices vastly higher than what they had paid only a few years before.

These giant canals were mighty works of engineering, the likes of which had not been seen before on the Rio Grande. In 1896 the normally matter-of-fact U.S. government engineer William W. Follett somewhat derisively referred to them as "monster canals" and to the land-sales promotions that accompanied them as "bonanza farming."[8] Monster they were; the Rio Grande Canal was ninety feet wide at the land surface, sixty feet wide at full depth, and contained a six-foot depth of water when filled to capacity.[9] The combined capacity of the four largest canals mentioned above (and there were many smaller ones) was about 3,500 cubic feet per second. For comparison, the mean summer flow of the Rio Grande at Del Norte is only 1,500 cubic feet per second.[10] Even if the water was not needed for irrigation in the San Luis Valley, it was diverted anyway. The unneeded water was dumped down the local creeks where some of it overflowed to irrigate meadows.[11] The residual ultimately drained into the Closed Basin, the bed of the ancient lake that once occupied the valley, where it evaporated.

In less than ten years, the newcomers had built diversions capable of literally sucking the Rio Grande dry at its source. This quickly came to pass. In 1879 about 122,000 acres were irrigated in the San Luis Valley. By the time the era of "bonanza farming" came to an end in 1892, more than 400,000 acres were under irrigation.[12] These consumed a flow of about 1,000 cubic feet per second, or two-thirds of the flow of the Rio Grande as it entered the San Luis Valley (see plate 13).[13] This water was diverted without the slightest consideration for the native farmers downstream who depended on the river for their livelihood and whose ancestors had been irrigating from the river for hundreds of years. Previously, the human populations along the banks of the Rio Grande had been able to make a dent in its flow only during the summer drought period. Now modern engineering technology had been unleashed at its headwaters, and most of the water in the main stem was siphoned off, regardless of whether it was needed. Neither the ecosystem associated with the river nor the farmers downstream would ever fully recover.

Throughout the 1880s the diversion of Rio Grande water in the San Luis Valley steadily increased. Farmers downstream in New Mexico and Texas might have received early warning of the growing crisis, but nature conspired to keep them in ignorance. Between 1880 and 1886, the San Juan Mountains received unusually heavy snowfall, and even the monster ditches were not capable of diverting the torrent of spring runoff that ran past their headgates and down to New Mexico. During 1887 and 1888, there was average snowfall in the mountains, but abundant summer thunderstorms in New Mexico kept enough water in the river to supply irrigation needs. Beginning in 1889, however, many winters produced only average-to-low snowpacks in the San Juans, and the summers yielded only weak summer rains. Under these conditions, the recently arrived irrigators in the San Luis Valley sucked the river dry. Without downstream replenishment from summer runoff, the river ceased to flow for months at a time.[14]

The dreams and visions of the new settlers from the East had come true. Unlike the busted homesteaders who had expected the rains to follow their plows, the entrepreneurs and farmers who had debarked from the railroad in the San Luis Valley succeeded in making the desert bloom. They established a thriving agricultural community where before there had only been sagebrush and sand. But this success was not free; a steep price was paid, not by the new settlers, but by the thousands of subsistence farmers downstream for whom flowing water in the Rio Grande was the foundation of their existence.

Drowning During Drought

Paradoxically, while the summer flows in the Rio Grande dried to a trickle, the farmers along its banks experienced both an increase in devastating floods and the gradual inundation of their farmland by pools of salty water.

The combination of the scourges of too little water and too much water must have seemed inexplicable to the subsistence farmers who suffered under both, but in hindsight they are the logical outcomes of heedless use of a delicately balanced natural system.

The link between too little water and too much water was the sediment that was delivered by the surrounding landscape; more specifically, it was related to the river's ability to carry that sediment. The amount a river can carry is determined by a combination of the amount of river flow and its speed.[15] The river constantly interacts with the land over which it flows. If the landscape supplies less sediment than what the river is capable of carrying, it will erode its bed to supply the missing sediment and, in doing so, will cut a canyon or arroyo. Conversely, if the landscape supplies more sediment than the river can carry, the sediment will drop out of the water and the bed of the river will rise.[16] Geologists refer to this deposition of sediment in and along the riverbed as *aggradation.*

This is what happened to the Rio Grande in the late nineteenth and early twentieth centuries. The river suffered two blows to the stability it had enjoyed for centuries. The first was a large increase in the amount of sediment washed in, due to erosion caused by overgrazing and clear-cutting of timber. The second was the decrease in flow caused by the diversions into the new monster ditches in the San Luis Valley. Either one of these would have caused the riverbed to aggrade; together, they caused rapid aggradation, disastrous for New Mexico farmers.

Rivers naturally regulate their own flow. When a river rises in flood stage, the sediment load in the fast-flowing water increases. If the river overflows, the water moves outward over the banks, its speed slows greatly, and it drops much of the sediment it is carrying. This builds up the bank and produces a natural ridge of sediment that tends to keep the river from overflowing any further. These raised banks are natural levees, and when the river is in equilibrium, they improve the ability of the river to carry floods without overflowing. However, when a river is aggrading, the natural levees produce a dangerously unstable situation. The levees rise along with the bed of the river. Eventually there comes a flood large enough to push big flows over the levees. The floodwater then runs rapidly down the outer slopes of the levees and begins to erode them. Once a breach has formed in the levee, the entire river empties into the floodplain (since its bed is now higher than the floodplain), and the river establishes a new course.[17]

The pueblos along the Rio Grande had always been subject to occasional flooding, especially from intense summer thunderstorms, but beginning in the mid-1880s, the valley began to experience intense flooding and damage on a regional scale.[18] In the early summer of 1884, flooding damaged nearly every village between Albuquerque and El Paso, Texas. Several towns were so devastated they were permanently abandoned. Very similar floods struck the next summer. In September 1886 severe flooding struck the entire Rio Grande from south of Albuquerque to Mesilla, New Mexico. The little town of Bowling Green near Socorro, New Mexico, was wiped off the map. Slightly smaller floods struck in 1889 and 1891. A lull followed, but in May 1897, the Rio Grande rampaged along its entire course through New Mexico, reaching a peak discharge of 21,750 cubic feet per second at San Marcial, New Mexico. Over the first thirty years of the twentieth century, eleven more regional floods caused extensive damage, culminating in the catastrophic flood of 1929 that obliterated San Marcial.[19]

The Saga of San Marcial

In 1902 a visitor to the town of San Marcial, New Mexico, south of Socorro (see plate 2), would have enjoyed its beauty. As described in the *Socorro Chieftain* that year, San Marcial was "one of the garden spots of the Rio Grande valley. Its avenues are lined with rows of large cottonwood trees. [There is] a beautiful park, in which a fountain and bandstand are located."[1] The town of about eight hundred residents was the second-largest in Socorro County.[2] It housed a railroad maintenance yard, freight depots, a Harvey House (one of a chain of railway hotels and restaurants built by the Fred Harvey Company throughout the West to support the burgeoning railroad system), a newspaper, and numerous mercantile establishments. In every stretch of the Rio Grande with good alluvial land on the valley bottom, there was a hamlet or little village: Valverde and La Mesa above San Marcial; San Geronimo and San Elezario close by; and Contadero, Cantarecio ("the place where the river sings loudly"), and Paraje about ten miles downstream. Each of these little towns had its own acequia and surrounding green fields. The area had a vigorous economy in which the local agriculture was supported by outside technology and trade arising from the railroad. Each village had, apparently, a bright future.[3]

Today, a visitor to the same spot would see only dense and uninhabited stands of mosquito-infested saltcedar interspersed with empty fields of salt grass, broken up by occasional brackish marshes. This almost impenetrable thicket carpets the valley bottom all the way to the delta of Elephant Butte Reservoir several miles below, where it is replaced by an expanse of dead saltcedar that perished by drowning when the reservoir filled. What can explain this dismal transformation?

San Marcial at the turn of the century, showing Main Street looking north. New Mexico State (University Library, Archives and Special Collections) Image 02231130.

The history of the San Marcial area has much in common with the earlier story of Chaco culture. In both cases a flourishing society pushed too far against the limits of its environment and suddenly collapsed. In the case of San Marcial, the town was wiped off the face of the earth in a little more than a month. It was as though the penalties arising from all of man's abuse of the Rio Grande were focused on one hapless community. The modern ambitions for better transportation, better irrigation, and a merchant economy conspired to convert a prosperous valley into a desolate swamp.

San Marcial was initially settled by Hispanic farmers around 1866. Many of them were refugees from a nearby flood that year, and they sited their little village mindfully on a raised bench at the west edge of the valley, a little above the apparent flood line. The seeds of San Marcial's fate, however, were sown about twenty years later when the Atchison, Topeka, and Santa Fe Railroad decided it needed an additional major maintenance yard well south of the existing one in Albuquerque. Instead of respecting the wisdom of the earlier settlers and locating their facility on higher ground, they carefully platted a new town in the valley bottom, along the railroad tracks.[4]

As recounted above, the entrance of the railroad only a few years earlier wrought huge changes in the landscape. The new immigrants to the San Luis Valley began their frenzied push to divert as much of the Rio Grande's headwaters as they could. Spurred by the new markets, vast herds of livestock denuded the watersheds while logging companies clear-cut the mountain forests. The Rio Grande responded to the decreased flow and increased sediment load by dropping the sediment on its bed. The river had a long history of destructive floods, but the rising bed now worsened this problem, which climaxed at unfortunate San Marcial.

By 1914 even under normal conditions, the water level in the river was two feet above the streets of the town.[5] The railroad began a long fight against the river, building levees and riprapping them, then raising the levees as the riverbed continued to aggrade. The railroad tracks ran along the east side of the town, helping to buttress it from the river, but about three-quarters of a mile below the town, the railroad crossed the Rio Grande on a bridge, leaving the municipal area unprotected from that direction. It became necessary to build an additional levee from the bridge to the hills west of town in order to prevent downstream floodwaters from backing up into the town. Until 1929 the railroad was successful in keeping major floods out of San Marcial, but by that time, the little towns downstream were gone.

After the San Luis Valley diversions had begun in the 1880s, the unreliable flows of the Rio Grande created intense demand for a large storage reservoir. That demand was eventually met when the new U.S. Bureau of Reclamation started construction of Elephant Butte Dam in 1913. The elevation of the dam was limited to avoid flooding San Marcial, but the little communities below it became a sacrifice zone. All of the private property was condemned and the inhabitants forced out. Most of the residents unhappily left the valley, but a hundred or so defiant ones held out until their land was finally inundated when Elephant Butte Reservoir completely filled in 1924.

San Marcial met its fate soon after. By 1929 the river had risen about fourteen feet above its original elevation, and the riverbed was now three feet above the town.[6] Disaster was not a question of *if*, but rather *when*, and the answer came with the summer monsoon of 1929. Intense pulses of atmospheric moisture boiled up from the south and, on August 10, erupted into downpours in the hills to the west. The rains continued for another three days. Torrents poured out of the arroyos to the west of the town. The low east and south sides of the town had been ringed with levees to protect against the rampages of the Rio Grande, but now these only exacerbated the flooding as the waters flowing in from the highlands to the west backed up against them. The town's flood protection acted as a bathtub from which the flood waters could not escape.

A decision was made to breach the dikes to the south to let the floodwater out. The decision was a calculated risk and it failed resoundingly. The monsoon outburst had continued to push northward, and the runoff from the intense storms roared southward in a flood wave down the Rio Grande. The peak of the wave reached San Marcial on August 13, shortly after the southern levee had been breached, and poured through the gap. The little town was flooded to a depth of eight feet. The railroad telegraph proved a lifesaver when operators upstream telegraphed word of the oncoming flood wave and the AT&SF sent emergency trains to evacuate the inhabitants of the town just barely before the flood hit.[7] Although no lives were lost, the town was devastated.

It might seem that San Marcial had paid a full price, but the Rio Grande had still worse in store. Another monsoon outburst surged northward from September 21–23, hitting hardest in the Rio Puerco and Rio Salado watersheds to the west of the Rio Grande valley. The runoff pulses from vast areas were funneled through the narrow outlets of these two rivers into the Rio Grande, where they coalesced and surged down the valley in a flood wave even higher than the one in August. At five thirty in the afternoon on September 24, the river burst through the weakened levee above San Marcial and once again inundated the town. Almost as if the river were determined to forever obliterate San Marcial, this flood wave came armed with an additional weapon in the form of sediment. The floodwaters looked like churning red-brown syrup. The syrup was the topsoil of the Rio Puerco drainage, laid bare by some of the worst overgrazing in the state and now stripped off by the millions of tons under the intense rain. In the basin created by the levees, the waters were relatively still, and the flood dumped its load of sediment, sealing what was left of the town under three feet of silt.

The floodwaters ponded up behind the levees, and San Marcial remained inundated until the spring. No one, neither the railroad nor the private citizens, made any attempt to reestablish the town. The railroad yard operations were moved to Belen, New Mexico. The original Hispanic village of Old San Marcial was much less damaged than the newer railroad town,

Photo taken from the water tower of New Town, San Marcial, a month after the 1929 flood. The water-soaked buildings eventually collapsed. Photo courtesy of Socorro County Historical Society.

The San Marcial Opera House was destroyed in the 1929 flood. Photo courtesy of
Socorro County Historical Society.

and many inhabitants hung on, but another flood in 1937 wiped out most
of what was left. A family named Gonzalez moved to Socorro in 1952.[8] They
were the last inhabitants of Old San Marcial and, in fact, the last inhabitants
of the entire valley.

Today the valley is ruined beyond recovery. The final blow was delivered,
not by the raging Rio Grande, but by small trees. After the small farmers
deserted the valley, invasive saltcedar trees gradually crept over the abandoned fields. Today, they form a dense, impenetrable thicket beneath whose
canopy is only bare, salt-encrusted soil. Saltcedar is not native to New
Mexico, but evolved over millions of years to prosper in the saline delta of
the Euphrates River in Iraq. Unlike most plants, the saltcedar pulls brine
from the soil and moves it up to the leaves. Ordinary plants would die as
the salt began to accumulate in the leaves, but the saltcedar has developed
special organs that remove the salt and excrete it as tiny drops of brine
that drift down from the leaves to the soil beneath. The saltcedar is thus
actually a pump, continually extracting salt from the shallow groundwater
and depositing it on the soil.[9] Over the eighty years since the abandonment
of the valley, this salt has built up in the soil to the extent that, were some
miraculous plague to eliminate the saltcedars, over much of the area the
native cottonwood and grass could not regrow because the soil is now too
saline.

According to the book of Genesis, the Vale of Siddim, at the south end
of the Dead Sea, was once "an irrigated garden, like the valley of the Nile"
that supported five small cities.[10] Today, locals take visitors to "Lot's Wife," a

pillar of salt that keeps watch over a barren plain of salt and sand. According to Genesis, Lot's wife bears witness to God raining fire and brimstone on its sinful cities. However, one does not need to travel to the Middle East to see such a spectacle. Simply take Interstate 25 in New Mexico to exit 124 and drive east to the bluffs overlooking the site of San Marcial. The land you will see before you used to be the *Val Verde* (Green Valley) where a sandy river meandered between cottonwood groves past a prosperous little town and its outlying villages and scattered green fields. Today the valley is uninhabited, and all the towns and farms are buried deep beneath the muck. The very soil is poisoned so that it can only support a jungle of noxious saltcedar. The Val Verde bears its own silent testimony, not to the wrath of God, but to the foolishness of man.

Notes
1. M. Marshall and Walt 1984, 283.
2. Ibid., 287.
3. Ibid., 265.
4. Gault 1914, 3.
5. Burkholder 1928, 129.
6. Harden 2006.
7. Ibid.
8. Anderson and Barrows 1998.
9. Glenn and Nagler 2005.
10. Gen. 13:10.

Many factors contribute to repeated flooding of this nature, including fluctuations in climate and weather, but two stood out for the Rio Grande in the late nineteenth and early twentieth centuries. The first was the propensity of the river to change its course frequently due to the aggradation of its bed, as already discussed. The second was the tendency for floods to produce about the same amount of runoff as previously, but in a shorter time, known as *flashiness*. This flashiness was largely due to the stripping of vegetation through overgrazing and clear-cutting. Before, plant leaves and branches had slowed the flow of rainfall to the land surface, and the leaf litter and porous soil beneath the plants soaked up rainfall like a sponge and slowly released it to the streams. With the vegetation stripped, the rain could directly hit the hard-packed soil and cascade down the bare slopes, producing peak flows much greater than before the natural system was exploited.[20]

The combination of greater flood peaks and aggraded channels was devastating to the long-established balance of the river on which subsistence farming relied. When these raised channels were hit with huge floods running off the denuded landscape, the river quickly cut through the natural levees and the entire river took a new course—through fields and even entire towns—covering the valley bottom with water for months.

Salt of the Earth

Another type of flooding was also becoming common. This was more insidious and could be traced to the same causes. Water gradually seeped out of the soil, coalescing into ponds and swamps. These swamps and the soil around them became coated with a white crust of salt. A farmer in the north valley of Albuquerque recalled that, in the 1920s, "Alkali covered the land on Rio Grande Boulevard from Pueblo Road clear to Chavez Road. You see it was white like snow. Everywhere was water—there were lakes all along the land. There were no houses there—you couldn't even walk there. . . . The land wasn't any good at all until they dried it."[21]

From the eighteenth to the end of the nineteenth century, much of this land had been productive, irrigated by acequias. However, the traditional water distribution system carried the seeds of its own demise. Modern irrigation canals now end in channels called *wasteways* that run downslope into a lower canal or a drain, instead of contouring the slope as an acequia does. These wasteways convey any excess irrigation water out of the area. Traditional acequias, in contrast, simply ended at the last farm on the ditch. The excess water was released to soak into the soil.[22] In the short run, this practice produced *cienegas* (swampy meadows) that yielded grazing for livestock. Over the longer term, however, the effect was to raise the water table of the whole area. As it approached the surface, the groundwater was wicked upward and evaporated, leaving behind the salts dissolved in the water. The water in the Rio Grande near Albuquerque contains about 175 milligrams of salt per liter of water.[23] To put this in perspective, if an Olympic-size swimming pool were filled with this water, about 1,000 pounds of salt would be left behind after the water evaporated.

The rising water table and accumulating salts were a constant threat, but until the late 1800s, there was some outlet for them. When flow was low in the Rio Grande, the river's surface was below that of the water table. So the groundwater slowly flowed toward the river and seeped into the channel, acting like a release valve on a pressure cooker. This release valve was plugged shut when the river began to aggrade its bed in the 1880s. As the riverbed rose, the channel became higher than the surrounding fields, and the flow direction reversed. Water now constantly seeped out of the channel and flowed outward, raising the water table until it broke the land surface.

Now there was no outlet for the salts. The salt that had accumulated over centuries of acequia irrigation and the new salt carried in by the river seepage began to crust the soil, producing fields "white like snow" and useless for agriculture.[24] Many were prospering in the new Rio Grande basin, transformed by the advent of the railroads. The railroad barons amassed fortunes; the timber barons stripped the forests and then left the area; and the cattle barons shipped out animals by the millions and reaped the profits flowing back. The size of the middle class surged, building on a great expansion of commercial activity. Anglo farmers with access to capital bought up land in areas fortunate enough to escape flooding and salinization and established successful farms, which supplied local and distant markets. However, the indigenous inhabitants of the Rio Grande valley, both Hispanic and Indian, were caught in a vise. They relied on their small plots of land for their food supply and other basic necessities, but between lack of irrigation water and encroaching salty swamps, they were increasingly unable to produce successful harvests. Frequent floods damaged their fields and dwellings, but without income from crops, they were unable to make repairs. Large-scale infrastructure such as hardened levees and deep drains might have solved their problems, but they did not have the capability to construct such massive works themselves, and, without a cash economy, they could not pay others to construct them. The Rio Grande had been transformed from their ancient friend, the lifeblood of the region, to an unstable and dangerous enemy.

Chapter VI
Conquest of the River by Science and Law

Prior to the adoption by the New Mexico Territorial Legislature of a 1907 water code, a new perspective on human uses of natural water gradually emerged. This new perspective had three foundations: the belief that modern *science* could fully comprehend the mysterious behavior of western rivers like the Rio Grande; confidence that modern *engineering* could use that understanding to control the rivers for human benefit; and faith that modern *law*, based on the best of science and engineering, could successfully administer the rivers that were brought under control. Underlying and supporting these three pillars of water policy was the unquestioned presupposition that employing the best of all three met a societal obligation to dominate nature and make it better serve the needs of men.

And so, near the end of the nineteenth century, the new U.S. Geological Survey (USGS) set out to apply science to eliminate the mystery of the Rio Grande by quantifying the flow of the river. Soon thereafter, the U.S. Reclamation Service set out to engineer away the river's natural inefficiency. Establishing a modern, rationalized law code would prove more contentious. The old water laws were idiosyncratic, inconsistent, and very local. The practices were not easily amenable to conforming to a single rule of law that gave equal weight to the interests of private individuals. Nevertheless, by 1907 these new values were embodied in a formalized law, centralized at the state level. Raising these three pillars of a modern hydraulic society on the Rio Grande was a formidable challenge, but the issues were broached, the effort was undertaken, and the outcome became assured in the twenty years between 1887 and 1907.

Hydrologic Science to the Rescue

While the native and Hispanic inhabitants of the Rio Grande valley were baffled and frustrated by the changes in the river, the recent immigrants looked to the newly emerging powers of engineering and science to resolve the developing problems of flooding, erratic river flows, and salinity, and even expand uses of the river for the new wave of farmers who were headed west to make the desert bloom. The first order of business was to determine how much water was actually available. How big was the pie to be divided up between Colorado, New Mexico, Texas, and Mexico? How big were the bad floods? What size levees would be needed to restrain them? Most important, how much unused water could be applied to new development? In order to answer these questions, the flow of the river would have to be gauged.

There was some small precedent for measuring flows of water in the area. Hispanic New Mexicans had experimented with the *surco* as a basic measurement of water flow, but had never agreed on what the surco actually measured. (The values ranged from one cubic foot per second, to the rate of flow through a ditch filled, to the level of a wagon wheel.)[1] Even if they could have agreed on the surco, they had no way of determining how many surcos flowed past them down the river. The nascent USGS began to address this lack of main-stem flow information in December 1889, when it established the first stream-gauging station in the United States. The gauge was on the Rio Grande near Embudo, New Mexico, one of the railroad stations on the Chili Line between Santa Fe and Taos.

The initial gauging work was conducted as a kind of training school for junior engineers and hydrologists who went on to lay the foundations for the entire national gauging network of the USGS.[2] The training was conducted under harsh conditions. Like many later tourists, the student gaugers were surprised to find that this part of New Mexico was not balmy desert; at an elevation of six thousand feet the gloomy depths of the Rio Grande gorge were windy and bitterly cold. The tents and blankets provided were so inadequate that some of the trainees dug a small cave in which they were fairly comfortable until a wandering goat fell down the chimney and collapsed the dugout.[3] Their transport for gauging the river was a tiny raft consisting of four small barrels nailed together with planks; it was conveyed across the river by ropes precariously tied on either bank.

At the head of these junior engineers was twenty-seven-year-old Fredrick H. Newell. A savvy and ambitious young man, he had only the previous year attained the rank of assistant hydraulic engineer under Major John Wesley Powell, the director of the USGS.[4] This was a signal accomplishment for someone only three years past graduation from the Massachusetts Institute of Technology, but as he leaned over the slippery edge of his tiny, icy raft being winched across the partially frozen Rio Grande gorge in winter 1889, he must have wondered if he had made a mistake by not taking a position in a comfortable office back east.

At each stop in the slow journey across the river, Newell or one of his trainees let out on the end of a rope an object that resembled a tiny torpedo, recorded some numbers, pulled it in, and then did the same thing a little further across the river. The little device they dropped in the river was a current meter, one weapon in the arsenal of new technology brought to bear on the wild rivers of the West.[5] The propeller on the aft end of the meter turned faster as the current grew swifter; by recording the rate of revolution the young engineers could compute the cubic feet of water flowing down the river each second.[6] Two young men at work on a raft in the middle of an icy river might seem innocuous, but each rotation of the propeller brought closer the day when the Rio Grande would be brought under control behind dams and levees.

Student hydrogeographers in the Embudo stream-gauging class. Fredrick H. Newell is in the top row, near the center, gazing outward. Photo courtesy of U.S. Geological Survey.

The Embudo gauging crew arrived in December 1889 and left less than six months later. These newly trained stream gaugers then dispersed north to Colorado and west to California to measure other big undeveloped rivers. They had measured the base flow of the Rio Grande from December to April, when the river runs relatively low and steady while the surrounding mountains in the basin accumulated the winter snows. These Rio Grande base flows already were committed to preexisting uses. What the Embudo gaugers missed were the Rio Grande flows they were really after: the April to October flood flows, which raged erratically with melting snow and torrential summer rains. These flood flows were the terror of acequias and would become the dream and target of the new development promoted by the United States. But the Embudo gaugers had taken the first step toward transforming the Rio Grande into a quantifiable commodity that could be reliably delivered to those who claimed ownership over it.

Who did own the base flow of the Rio Grande, and how much of the flood flows would be left over for new development? It was a question that would gather more and more importance as the age of reining in the river unfolded. The federal government started by seeking a more comprehensive understanding of the nature and extent of existing claims to the river. Although the landscape and river had been changing rapidly throughout the late 1800s, due to exploitation by railroad, timber, and cattle barons, and by upstream farmers who benefited from the new diversions, ownership of the river itself had not been determined. Ancient acequia rights had been recognized early without defining them, but the legal system had never addressed who, if anyone, owned rights to the water in the main stem of the Rio Grande.

In 1897 the International Boundary and Water Commission (predecessor to the U.S. Reclamation Service and the Bureau of Reclamation) dispatched William W. Follett to the headwaters of the Rio Grande in the rapidly growing San Luis Valley of Colorado and told him to survey existing diversions of the mighty river from its headwaters to the twin cities of El Paso, Texas, and Ciudad Juarez, Mexico. His most pressing task was to figure out why the Rio Grande had recently begun drying up at Albuquerque and downstream during the summer, and what might be done about it. Traveling south mostly by wagon, Follett spent more than a month checking on existing points of diversion and getting estimates of acreage irrigated by each diversion from ditch bosses and mayordomos.[7]

This would be the first of many official efforts to create a comprehensive list of claims to the entire Rio Grande basin. Follett's rough estimates were astonishing in the breadth and depth of existing uses he recorded along the river. He estimated that perhaps 180,000 acres in the basin already had a history, sometimes centuries long, of irrigation. Follett laid the blame for the recent water shortages in New Mexico, El Paso, and Mexico squarely on diversions by the new irrigators in the San Luis Valley.

In spite of the mid- and late-summer shortages, much of the spring snowmelt pulse was not captured. Through careful consideration of the few years of gauging then available, he estimated that each year an average 400,000 acre-feet of water flowed past El Paso, unused by humans.[8] This water was flood flow that in the thinking of the time was "lost," because it could not be captured for irrigation. No consideration was given to the inherent value of the river itself or to the other biological communities that the river supported. The dreams of men like Follett focused on the additional irrigated acreage necessary to really make the desert bloom. If these exclusive dreams were ever to be realized, the flood flows of the river, then alternately terrorizing and starving existing users, would have to be captured and used. That would require massive storage. Storage would require tall dams. Tall dams would require a firm enough legal structure of water rights to support the huge investment to build them.

The legal structure for harnessing the unappropriated flood flows, Follett noted, was lacking. As of 1896 Follett reported there was no comprehensive Rio Grande water law as far as he could see. He observed the ditches where they connected to the river, but nothing more. He noted that in the middle Rio Grande, where he found almost eighty community ditches irrigating perhaps 140,000 acres, ditch bosses complained constantly about excessive upstream uses of the river but had no legal recourse for resolving the problem. What Follett was looking for was a comprehensive, rational, legal system that controlled the system as a whole. The governance of the diverse and intricate acequia systems was irrelevant to him; his focus was broader than that. Beyond each ditch, the river still ran free, not much affected by the modest, unsophisticated acequias. He believed only a comprehensive law of the river could ensure public development of the resource.

Getting Legal Control of the River

Throughout the 1870s, when acequia irrigation reached its peak in the Rio Grande basin, the extension of physical and legal controls into the great river had been incidental, sporadic, and unintentional. Territorial courts of general jurisdiction arose for the first time in territorial New Mexico. Although acequia affairs were generally left to the acequias themselves or to the new, very local justices of the peace, federal district courts of the Territory of New Mexico also occasionally got involved. As early as 1861, one territorial district court divided the shared resources of a Rio Grande tributary among competing acequias. The court apportioned the streamflow by time, allotting to each ditch a set number of days over a specified rotation period for the exclusive use of the ditch. The decree solved a practical problem of competition for a shared resource, but most important, it did so by extending a time-honored acequia tradition—*por tandas* (by turns) distribution of water within a ditch—to the river itself. In the second half of the nineteenth century, other New Mexico courts followed suit.[9]

The new territorial legislature also sporadically and without an over-all plan extended public controls from the acequias to the rivers from 1854 to 1890. The Territorial Legislature became perhaps the primary arbiter of local claims to shared sources in the Rio Grande basin. For example, the legislature quietly declared that the Rio Grande del Rancho tributary to the Rio Grande was itself an acequia, to be governed jointly by two competing acequias. Just as the U.S. courts had extended acequia apportionment proce-dures to the rivers, the territorial legislature figuratively made certain rivers ditches. U.S. law thus started on a path it would pursue single-mindedly in the mid-twentieth century: to turn the Rio Grande into the equivalent of an irrigation ditch.[10]

To free the waters of the Rio Grande for new uses, the new territorial government had to gain formal control of the long-established informal claims to the river. In cooperation with like-minded politicians in and out of New Mexico, the Territorial Legislature began to create legal identities for traditional New Mexican water users. In 1895, in a little-noted but highly significant move, the Territorial Legislature declared the New Mexico pueb-los and acequias, the two primary groups with existing claims in the Rio Grande basin, to be corporations.[11] Thereafter, the courts referred to them as "public involuntary, quasi corporations."[12] Their new identity transformed them from informal, local, and communal entities that had dealt with the river for centuries to a status that was formal, legal, and state conferred. As corporations, the pueblos and acequias now had the power and necessary legal identity to enter the world of broad state control of the river they shared and could formally sue and be sued.

And sued they were, mostly by one another.[13] The effort to bring New Mexico's traditional water institutions into the twentieth century by giving them twentieth-century identities merely freed the pueblos and acequias to fight each other over their relative water rights. In the early 1890s, it looked as if the territorial courts, through case-by-case litigation, might be the ulti-mate arbiters of the relative claims of the pueblos and acequias to the Rio Grande flows.

Competing Claims

In the absence of a legislative law of the river, the Rio Grande was subject to all manner of claims in the last decade of the nineteenth century. At one extreme were international claims from downstream Mexico; in the middle were conflicting state and federal claims to control the river; and at the other extreme were private claims that threatened all public control. Each claim cried out in its own way for a general, consistent, and uniform framework for the Rio Grande. The process of resolving these individual claims from 1872 to 1907 initiated the construction of this framework.

In the 1890s Mexican complaints about the international management of the Rio Grande became increasingly strident, because each year less Rio Grande water reached Ciudad Juarez and the irrigated lands around it. The U.S. State Department became embroiled in the Mexican complaints, and ambassadors from both countries found themselves in the unenviable and unusual position of Rio Grande water negotiators.

Within the United States a crucial question came to the fore: who would control the anticipated and yearned-for new development of Rio Grande water: the federal government, the territorial government, or private enterprise?

This proved a controversial topic. One unique answer was supplied by John Wesley Powell in his famous 1879 report on the arid lands of the West.[14] Powell, who knew the western rivers firsthand and better than almost anyone, argued that prospects for increased irrigation in places like the Rio Grande basin were really quite limited, dampening the wild dreams of speculators. Of equal importance, Powell also argued for the creation of watershed-wide federal water districts that might even cut across state lines. But Powell did not prevail. Had his basic proposal succeeded, a single Rio Grande federal district might have spanned those parts of Colorado, New Mexico, and Texas that were within the basin. Powell's solution to western water would have headed off the impending conflicts between Rio Grande basin states, the federal government, and the larger-than-life private speculators who proposed to develop the basin themselves. The Powell proposal, in essence, started a ten-year war in Congress over water in the West, which pitted Powell and federal bureaucrats who favored national control of water development against people such as Senator William ("Big Bill") Stewart of Nevada, a powerful proponent of state and private development of western water.[15] In the water debates of the 1890s, Stewart, a Harvard Law School graduate with a deep understanding of western resource issues, took to packing a six-gun around the Senate to prove to his eastern colleagues the truth of the western saying that whiskey was for drinking and water for fighting.

The water war ended in a draw with passage of the new Reclamation Act of 1902.[16] The act gave the federal government control over the planning, financing, and construction of the massive new dams, rather than handing them over to private interests. But the act also explicitly stipulated that water for new development would be governed by state, not federal law, and that New Mexico, not the United States, would apportion the rights to water for its new projects.[17] The United States would retain its indirect constitutional powers over western water: to regulate interstate commerce, to control navigability on streams, to control federal lands, and other powers. But the basic power to define and create rights to Rio Grande waters was ceded to the states and territories.

The Great Dam Controversy

On May 21, 1903, an empire slipped between the fingers of Dr. Nathan E. Boyd.[18] He had come remarkably close to establishing himself as virtual king over the Mesilla and El Paso valleys, but on this day the New Mexico Territorial District Court stripped Boyd's Rio Grande Dam and Irrigation Company, of which he was the principal mover, of its crucial right-of-way to construct a dam across the Rio Grande narrows at Elephant Butte. The grounds for the forfeiture were that the company had failed to begin construction of the dam within the five-year limit specified by federal law. This justification galled Boyd to the day of his death, for the actual causes of the delay in dam construction were administrative and legal obstacles thrown up by the federal government itself.

Behind these maneuverings lay a profound shift in the political philosophy of development of the American West. Once it was clear that new rights to water in the Rio Grande would originate from the states or territories and not the federal government, the question quickly became whether some public entity would also finance and build the new, massive infrastructure that the development would require or whether private enterprise would raise the capital, do the work, and in the end control the water itself. In New Mexico, this issue ultimately revolved around the site of a dam on the Rio Grande in southern New Mexico, with Boyd leading the debate for private development.

Boyd was not an engineer, but a doctor from Virginia, who was also a flamboyant operator, an international businessman, and clearly a man ready to seize an opportunity. In the expansionistic early days of the Gay Nineties, the capitalist viewpoint of Senator Stewart and his allies complemented the popular view that major infrastructure projects, such as dams, should be financed by private industry, and the government should keep its hands out.

Boyd immediately saw that whoever controlled a dam on the Rio Grande controlled the river, and whoever controlled the river controlled the economy. In 1893 he formed the Rio Grande Dam and Irrigation Company and rushed in an application for the right-of-way to construct a dam at Elephant Butte, the upper of the two locations fingered by Follett in 1896.[19] Follett's lower dam site, just above El Paso, was favored by the municipal governments of Ciudad Juarez and El Paso and was heavily promoted by General Anson Mills, one of El Paso's prominent citizens and a man whom Boyd considered a vicious political enemy. Follett had pointed out that there was water enough to construct only one dam, and Boyd knew that Elephant Butte held the trump card by virtue of being upstream. The right-of-way application was approved by the secretary of the interior in February 1895, and Boyd's grand private scheme was launched.[20]

New Mexico was then, as now, very poor in capital. Most entrepreneurs would see this as a major impediment to a giant project such as a dam at Elephant Butte, but Boyd realized he could turn this gap between capital and project to his great advantage. He went to England to search for investors and

Dr. Nathan Boyd poses with family: his daughter with black dog in lap, his wife with dog in lap, his future daughter-in-law (with glasses), and her sister. New Mexico State (University Library, Archives and Special Collections) Image no. 00010071.

relied on the vast distance between the two places to tell his different audiences different stories about what he intended.

In England, Boyd produced an advertising prospectus that told what he actually planned to do: demand exorbitant rates for the water and force the farmers to either go broke paying for it or else cede half of their land to him. He would require landowners below the dam to turn over half of their land to the company in exchange for the right to irrigate the other half. In addition, he would impose a $1.50 fee per year in perpetuity for every acre irrigated or force the landowners to purchase water rights at the company's going rate, essentially forcing large landowners in the area to sell out to him.[21] Boyd's pitch was successful; he raised $1.6 million from British investors for the project.[22]

In his New Mexico publicity, on the other hand, he stressed how his project would make available a greatly increased and much more reliable supply of water, an endeavor beneficent in all respects. He kept his plans for gaining control of everyone's land close to his vest.

For a while, Boyd's amazing scheme succeeded grandly. New Mexicans strongly opposed the El Paso Narrows dam site that was being heavily promoted by Ciudad Juarez and El Paso. New Mexican opposition was founded on the knowledge that all benefits would accrue downstream in Texas and Mexico, while the Mesilla Valley and Las Cruces, New Mexico, would still be left without a reliable water supply. Worse, the El Paso reservoir would flood 40,000 acres of Mesilla Valley farmland.[23] Consequently, New Mexico and the state's supporters in the federal government enthusiastically boosted Boyd's Elephant Butte Dam.

The flaw in Boyd's scheme was that he banked on the fact that he could tell one story in New Mexico and another in England. But someone leaked to New Mexico the contents of the Rio Grande Dam and Irrigation Company's prospectus that had been distributed in England. The reaction of the secretary of the interior, David R. Francis, when he was asked for a formal opinion on the Rio Grande Dam and Irrigation Company project, was typical: "Congress has never granted, nor authorized this Department to grant, to this company, or to any company, a monopoly of the entire flow of the waters of the Rio Grande, and power to reduce to servitude landowners, citizens of New Mexico, Texas, and Mexico, living on the river below the reservoir proposed."[24] New Mexicans realized that Boyd's project was a trap. There was only one other place to turn, and so the formerly despised proposal for a government-built dam at Elephant Butte began to take on a new luster.

The proposal to construct a public dam at Elephant Butte was strongly opposed by the government of Mexico, fearful that the already meager supply of irrigation water to the Ciudad Juarez area would be totally cut off by New Mexico. Boyd proposed that, in return for a subsidy from the U.S. federal government of $250,000 paid over twenty years, his company would be pleased to assure the Mexicans of a supply of irrigation water at the rate of $1.50 per acre-foot—water that the Mexicans had previously diverted for free.[25] This proposal only infuriated the Mexican government.

The hubris displayed in Boyd's riposte to the Mexican concerns contributed greatly to his undoing. The recently appointed secretary of state Richard Olney was anxious to maintain good relations with Mexico, and thus, when Boyd alienated the Mexicans, he tended to favor Mills's El Paso Narrows dam site. This in turn fueled Boyd's hatred of Mills until the federal cabinet members found themselves a secondary target of that antipathy. At one point, Boyd wrote to then secretary of state John Hay, "I shall take the law into my own hands, and by *horse-whipping* you in some public place thus assure the necessary publicity and discussion of the irrefragible [*sic*] proofs of Gen. Anson Mills' guilt and of the protection you have afforded him."[26] Furious with Boyd's arrogance, Olney attempted to use the federal government's jurisdiction over navigable waters to axe the Elephant Butte Dam. This case dragged on through numerous appeals in territorial and federal courts, twice rising to the U.S. Supreme Court. Olney was ultimately unsuccessful in convincing the courts that the Rio Grande was navigable, but his rear-guard action caused enough delay that events took a new course.[27]

The final undoing of Boyd's water grab was the Reclamation Act of 1902.[28] Teddy Roosevelt was then president, and under his leadership, public opinion shifted decisively toward federal planning and control of development for federal lands and state waters. Sponsored by Nevada congressman Francis G. Newlands, the bill provided for federal planning and construction of western irrigation projects, although the water itself remained subject to state laws. The New Mexico territorial government almost instantly switched sides and filed a plea demanding forfeiture of Rio Grande Dam and Irrigation Company claims. In April 1903, the federal government was able to file a supplemental complaint that the company had exceeded the five-year time limit for initiating construction. In only a little more than a month, the District Court of New Mexico ruled against the company in a decision that, though appealed numerous times, remained final.

Slight and seemingly arbitrary events had shaped an outcome that would determine much of the fate and even the physical shape of the Rio Grande and New Mexico's water future. Had Boyd been able to pull his case out of the courts a little earlier, had the Reclamation Act failed in Congress, had any number of other details gone differently, Nathan E. Boyd might have ruled over most of the irrigated farmland in southern New Mexico and west Texas like a feudal lord, a man of fantastic wealth and power. Instead, in May 1903 the court had abruptly undermined his grand scheme. The decision was upheld by both the Supreme Court of New Mexico and the U.S. Supreme Court. In the wake of Boyd's ruin, control of the future of the Rio Grande rested on an uneasy alliance between the governments of the United States and New Mexico. The wave of individuals who profited mightily from gaining control of natural resources had crested and receded. Governments moved in to control the water.

A Major Coup for the Young Reclamation Corps

The 1904 National Irrigation Congress was no ordinary annual meeting. Convened in El Paso on Tuesday, November 15, the congregation was the first to be held along the Rio Grande, the first to focus on the river, and the first to include the Reclamation Service, which had been established in 1902. This was actually the twelfth annual meeting of the group (the sixteenth would be held upstream in Albuquerque; see plate 14), which brought together prominent state and federal politicians, irrigation engineers, educators, developers, and other leaders of public opinion. By 1904 the group had emerged as a powerful force for federal involvement in reclamation in the western states.

Competing interests to the Rio Grande had by now fractured the debate over control of the Rio Grande waters. Private entities like Nathan Boyd's Rio Grande Dam and Irrigation Company competed with governmental entities like the United States over control of dam sites on the river. The states of New Mexico and Texas fought over their relative rights to the river. The federal International Boundary and Water Commission battled with the new Reclamation Service. Mexico complained bitterly that its relative international rights to the river were not being respected. All of these divisions were packed into a roaring debate about whether the Rio Grande needed a public international dam near El Paso or a larger multipurpose dam 125 miles upstream, near the site of Boyd's private debacle. In 1904 it looked as if no one, least of all the large National Irrigation Congress, could put the Rio Grande back together again.

Then B. M. Hall, a young Reclamation engineer, showed up in El Paso, having been dispatched by Chief Engineer Frederick H. Newell. Miraculously, over the four days of the conference, he was able to mold a Rio Grande plan from the previous chaos. Hall proposed an international tall dam at Elephant Butte, persuading both Texas and Mexico to abandon their insistence on a dam much nearer to the international border. He successfully convinced the New Mexicans, Texans, and Mexicans that such a dam would better suit the water needs of each entity. And he convinced them all that a federal project would be far preferable to anything with which a private entity or state could offer. Late in the afternoon of the meeting's final day, Friday, November 18, 1904, the full irrigation congress, almost four hundred strong, enthusiastically endorsed Hall's plan.

Adopting the use of religious language to reinforce the importance and sanctity of engineering projects, Hall later boasted:

> The results were more than satisfactory. They were nothing short of miraculous. It would have done your heart good to see those old sinners (the entities fighting over the Rio Grande) flocking to the mourner's bench. I had been doing missionary work among them for some time, but I had not expected such a landslide.[1]

This political support would give way to a flurry of governmental activity on the Rio Grande for the rest of the twentieth century. The dam, the political entities to support it, an interstate compact, and more impacts to the river would all follow. But the vast transformation began when Hall drove his solution to the river straight through the heart of the 1904 National Irrigation Congress.

Note
1. Littlefield 2008, 111.

The Bien Code Comes West

The Reclamation Service's annual report for 1902–1903, which surveyed the laws of the western states and territories, reached the following sweeping conclusion: "The [water] laws of many of the States and Territories are in a more or less chaotic condition."[29] The powerful official who delivered this pronouncement was none other than Frederick H. Newell, the engineer who as a young man had clung to an icy raft in the middle of the Rio Grande during the winter of 1889. At that point in his life, Newell was concerned only with determining how much water was carried by the Rio Grande, but now he was much more worried about how that flow was divided up.

However unpleasant Newell's job may have seemed from the comfortless perspective of his icy raft in 1889, his intelligence and deft footwork had enabled him to rise in little more than a decade to the position of commissioner of Reclamation, the kingpin of western water.[30] After being reassigned to Washington, D.C., in 1890, he had joined the Cosmos Club, the social home of Washington's elite engineers and scientists, and dined regularly with powerful men such as Gifford Pinchot, the future secretary of the interior. Pinchot introduced him to the governor of New York, Teddy Roosevelt, and Newell became Roosevelt's trusted adviser on western natural resources issues. When Francis Newlands pushed the Reclamation Act through Congress in 1902, with Roosevelt's strong backing, the president hand-picked Newell to head up Reclamation. From his new post high in Washington, D.C., Newell turned his vision back to the Rio Grande he had once known so intimately, in order to assess whether the existing water laws of the state would be capable of supporting the huge new projects he was going to set in motion.

In contrast to the preceding administrations, which had tended to favor large-scale capitalists, Newell shared with Roosevelt, Pinchot, and other prominent Americans associated with the new administration in Washington a passion for developing the West for the benefit of the commoner: in this case, the rugged, individualistic American farm family.[31] In 1918, Newell was awarded the Cullum Geographical Medal by the American Geographical Society. (The inscription on the gold medal read: "He carried water from a mountain wilderness to turn the waste places of the desert into homes for freemen."[32]) Newell's goal was not only to make the desert bloom, but also

to do it in such a way that the common man would reap the benefits. Boyd's attempted oligarchy over agriculture in the Mesilla and El Paso valleys highlighted the inadequacies of the current state laws in this regard.

Newell's assessment of state water law was necessary, because by the turn of the twentieth century, it was as clear as anything could be in complex Rio Grande matters that while the United States would control access

Frederick H. Newell, first chief engineer of the U.S. Reclamation Service and its second director. Photo courtesy of the Bureau of Reclamation.

The Progressive Movement and the Rio Grande

The late nineteenth century was known as the Gilded Age, a decade-long spree of unfettered capitalism that elevated a small group of the captains of industry to the lifestyle of kings. At least a little of the gilding on their palaces and yachts came from the Rio Grande basin. The profits from clear-cutting entire mountain ranges, from the huge herds of livestock that overgrazed public land, from the sale of land in enormous irrigation projects, and from the operation of the railroads that hauled all the components of these industries back and forth flowed into the pockets of financial magnates. Unfortunately, most of the environmental consequences of these endeavors were felt in the small communities along the Rio Grande whose residents paid the heaviest price and suffered from inadequate irrigation, floods, and salinization of their fields.

At the beginning of the twentieth century, the Progressive movement, exemplified by the policies of Theodore Roosevelt, arose in reaction to the excesses of the Gilded Age. The movement supported the ideals of participatory democracy, efficiency of government and society in general, government regulation of corporations and natural resources, and social justice.[1] An important part of the movement became identified under the phrase "conservation of natural resources."[2] When the Progressives spoke of *conservation* in terms of the Rio Grande, they were thinking in terms of an antidote to the environmental degradation that plagued the river. Central to this concept of conservation was efficient use of resources, such as reservoirs to store spring runoff, rather than the contemporary usage, which stresses preservation over efficiency.

Over the second half of the nineteenth century, a systematic study of the natural environment had been fostered in the universities of the eastern United States, especially at Yale.[3] Out of the gentlemanly field of natural history gradually emerged the modern professions of geology, hydrology, forestry, range management, and ecology. Following rigorous scientific investigation, these professionals began to recognize the finite limits of natural resources (e.g., the thousands of years required to form soil) and the interconnectedness of natural phenomena. They came to a disquieting realization that the rate and nature of contemporary exploitation was not sustainable.[4]

This realization led the federal government to assert an interest in managing natural resources. This policy was most notably brought to fruition by two individuals: Fredrick H. Newell, whom Roosevelt appointed to head the Reclamation Service in 1907; and Gifford Pinchot, who took over the Bureau of Forestry (predecessor to the Forest Service) in 1900.[5] Under Pinchot's leadership, the Forest Service regulated logging and severely restricted grazing on forest reserves. The policies enacted by these two men had a profound impact on the Rio Grande basin. The dams, drains, and modern water distribution systems constructed by Reclamation went far toward eliminating the scourges of dry irrigation ditches and salinized land. As Pinchot's regulations were enforced on the forested highlands of the Rio Grande, the torrent of silt from soil erosion gradually abated, and the Rio Grande slowed its cycle of aggradation, flooding, and seepage. Ironically, this activist government eventually drove out of business the same small farmers they purported to benefit. Rather than practicing land conservation to preserve the community, they were motivated by the ideal of building a better and different society through the efficient use of natural resources. The very day that Nevada congressman Francis G. Newlands's Reclamation Act was passed in 1902, Roosevelt wrote, "I want it [federal irrigation policy] conducted, so far as in our power to conduct it, on the highest plane not only of purpose but efficiency."[6] In 1908, Roosevelt proclaimed to the nation's governors, "[F]inally, let us remember that the conservation of natural resources . . . is yet but part of another and greater problem . . . the problem of national efficiency, the patriotic duty of insuring the safety and continuance of the Nation."[7] Massive public works were to be initiated for the benefit of forward-looking, efficient, and entrepreneurial farmers "who will return the original outlay in annual installments paid back into the reclamation fund; [who will use the land] for homes, and not for purposes of speculation or for the building up of large fortunes."[8] Professionally designed and industrial-scale water works, reliable water supplies, land drainage, forest management—all these things came at a price. It was a price that market farmers using the latest improvements in agricultural technology were quite willing to pay, but for the subsistence farmer trying just to grow enough corn and beans to support his family, they were crushing. The machine set in motion by Roosevelt, Newell, and Pinchot slowly replaced the old dominance of the natural-resources robber barons by rational, efficient, scientifically based management, but also incidentally squeezed out the centuries-old lifeways of many Rio Grande villagers.

Notes
1. Hays 1959; Stradling 2004.
2. Bates 1957.
3. Ibid., 32–33.
4. Howland 1921, 146–47.
5. Pinchot 1947; Brinkley 2009, 423.
6. Brinkley 2009, 424.
7. T. Roosevelt, 1908, quoted in Hays (1959, 125).
8. T. Roosevelt, quoted in Brinkley (2009, 663).

to dam sites and other land-related rights-of-way on federal lands, the individual states would control water to be developed by projects located on that federal land. Before the federal government would be willing to invest vast sums in water infrastructure, however, what was perceived as the chaotic law of the river on the Rio Grande had to be straightened out.

Newell knew that his vast water projects would not achieve their goal of supporting a society of prosperous small landholders in the irrigated West unless the water was apportioned by means of rational and equitable law. He also knew that the existing water laws of the western states were inadequate for the purpose. The task of the Reclamation Service would be greatly expedited if someone were to develop an improved set of water laws that would be acceptable on a regional basis. He found that someone in the person of Morris Bien.

Bien was an engineer, trained in the early, reputable water resources program at the University of California at Berkeley. He was also a Washington-trained lawyer. Astute observers would later use the term *enginoirs* to describe the most powerful western water administrators, hybrid creatures who were part engineer (which most of the administrators were) and part lawyer (which most of them formally were not). Bien was such a person, the embodiment of those twin drives to technical science and rational law that were about to transform the Rio Grande.[33]

Bien took his gospel from men such as Follett and Newell and presumed to start from a clean slate regarding western water law. At the eleven-day Second Conference of Engineers of the Reclamation Service (which started in El Paso in mid-November 1904 and concluded in Washington, D.C., in early January 1905), he proposed an irrigation code for the western United States consisting of seventy-three separate sections.[34] Bien took the European civil law codes as his model. Like the European codes, Bien's proposed code for western water aimed to be comprehensive, systematic, and exclusive, while what previously had passed for Rio Grande water law was diffuse, disorderly, and encompassed even Catholicism and family relations.[35] The final section of Bien's proposed code repealed "all other acts and parts of acts in conflict with this act" and thus tried to sweep away (unsuccessfully of course) more than four hundred years of complex, contradictory water history in New Mexico and the Rio Grande.[36]

Bien's draft code emphasized what would become the three basic principles of the western prior-appropriation doctrine. First, all water belonged to the public. It did not belong to the owners of the land above groundwater or the owners of the land through which surface water flowed, as the riparian water law of the eastern United States dictated. Bien rejected the latter approach, which would have left many potential independent and small-scale irrigation farmers working nonriparian lands without access to water. Boyd's scheme to use land ownership at critical riparian points along the Rio Grande in order to claim ownership of the river's water, and thus ultimately gain control of an entire regional economy, is the type of outcome Bien's code was intended to prevent.

Bien's code also emphasized a second basic principle: that private parties could acquire a private property right to use public water. The nature and extent of that private right would be based on the beneficial use of the public resource. The right would be limited to the amount of water actually used and not a drop more. There would be no water hoarding or water speculation as Boyd's private plan for Elephant Butte Dam had threatened.

Bien's beneficial use requirement thus introduced Rio Grande water law to the arcane science of determining exactly how much water any particular beneficial use consumed. In Bien's version, there would be no ranking of

Beneficial Use

The concept of *beneficial use* at the heart of the Bien Code is intuitive to those who live in water-short places. In arid and semiarid regions, water has always been simply too scarce to be squandered, hoarded, monopolized, or wasted. The notion that water ownership should reflect the amount that has been beneficially used simply honors the need to ensure that water is always productively employed. In Bien's hands, however, beneficial use went significantly further in its formalization of fixed property rights for the essential, fluid, and ever-changing resource of water. Under his code, beneficial use became, in the mantra of western water law, the "basis, the measure, and the limit" of the private right to use public water.[1]

Beneficial use became the *basis* of western water rights in its requirement that water be used for some human and economic purpose in order to establish a right to it. However, productive use did not necessarily equate with the value of the use—economic, ethical, or otherwise—of water. Instead, it was a throwback to the philosophy of Manifest Destiny: any drop of water that would reach the sea was considered wasted. Beneficial use basically equaled human use of any kind. The kind of use—domestic, irrigation, mining, or industrial—didn't matter as long as humans marshaled the natural resource and made it do something that nature alone could not produce. The exclusive emphasis on productive use spun off ancillary doctrines like the law of abandonment and forfeiture. If, for example, the owner of a water right established by beneficial use did not continue to use it, then the right would be lost.

The further requirements of the Bien Code for beneficial use to both *measure* and *limit* private water rights meant that the quantity of a water right was established by and restricted to the minimum amount of public water necessary for the use. This technical requirement that water not be squandered introduced a century of science devoted to the study of soils, crop growth, meteorology, and hydraulics, all designed to guarantee that a private water right would effectively consume no more public water than absolutely necessary.

This exclusive legal and scientific emphasis on human consumption, born of the Bien Code's emphasis on beneficial use, blocked the extension of the basic concept of water rights to sustaining wildlife or riparian habitats until late in the twentieth century.

Note
1. New Mexico Constitution, art. 16, sec. 2.

beneficial uses: all beneficial uses would be equal and scientifically limited to the amount of public water that each consumed.

The third element of Bien's proposed code—priority—made the time when the right was established the sole criterion in determining which right to a common source would prevail in times of shortages. At some times, but not all, enough public water would be available to supply the needs of all private rights to a single source such as a river. When water was not available to meet all needs, the senior user—the one who had the earliest established claim—would get all the available water for his beneficial use before the junior user got any. This provision was intended to encourage development of water resources by protecting investment in their development. An important aspect of the doctrine was that it also recognized the reality that flows in western rivers were highly variable, and it set up a mechanism to allot water during times of scarcity.

Priority in Time

Besides beneficial use, the Bien Code made *priority in time* the sole basis for apportioning varying public supplies among private claimants to it. In other words, the first person to establish a water right by beneficial use has a right to obtain the full amount of his water right from the available supplies, even if that means that everyone with a later right goes dry. Under the priority system, there is no sharing among senior appropriators (those who were first to establish their right to the water) and junior appropriators (those whose claims came later) during times of water shortage. The senior appropriator gets it all.

Two essential factors in the system seemed in 1907 to justify this harsh result of the priority doctrine. First, common surface water supplies varied from season to season and from year to year. At times there would be enough water for all and at other times there would not. While recognition of this variability required some method for apportioning short common supplies among those with a right to it, priority wasn't the only principle that might have addressed it.

It was the nature of beneficial use in the Bien Code that required short public supplies to go to the first appropriator. By definition, the first appropriator had invested capital to make the water do something that it would not have done naturally, and the investment had to be protected against a subsequent appropriator who might otherwise take the water and negate the earlier investment. To promote and protect beneficial use, the legal system had to protect the first beneficial user.

Unlike the principle of beneficial use, which has ancient roots, the strict principle of priority is of a much more modern and untested vintage. Perhaps for that reason, it is arguably clumsy in accomplishing the goals for which it was developed. For example, one of the unintended consequences of the doctrine of priority is that it tends to protect and lock in the earliest and least economically productive uses of common supplies. A senior farmer's right to water for alfalfa is, under the principle of prior appropriation, to be supplied fully at the expense of either a manufacturer's junior right to water for computer chips or to a growing city's claim to increased water for domestic use. This result seems at odds with the West's professed desire to grow economically through the most efficient use of water resources.

This arguable flaw in the experimental doctrine of prior appropriation makes it a clear example of the contradictory currents that flow through western water history. It was the protection of existing uses, rather than the promotion of new investments, that won priority protection a place in New Mexico's 1912 state constitution. The requirement of beneficial use always had been present in proposals for the new state's constitution. But the requirement that priority in time would confer the better right to water came in only at the last minute and at the insistence of Hispanics worried that the new state would somehow deprive them of access to common water sources. Thus, from the start, this fundamental concept of New Mexico water law—priority in time—raised contradictory economic and social justice concerns and tensions that continue today.

These three basic principles of Bien's code—public water, private rights to it based on beneficial use, and priority in time—were incomplete reflections of practices adopted in the previous centuries of intense water use in the Rio Grande basin. The public nature of water in the code was in harmony with the basic communal importance of water for the ancient water-based Pueblo and Hispanic communities. Similarly, the beneficial-use requirement of Bien's code sounded vaguely like the informal but exacting stewardship that ruled those earlier communities. And priority as the basis for apportioning short public supplies had part of its roots in the multifactored, subtle Spanish and Mexican system for dividing rights to public water. The notion of priority apportionment also had some precedent in the customary law of the California gold camps in the mid-1800s and in the earliest congressional declarations on western water from the 1860s and 1870s. But the Bien Code explicitly adopted the principle as a matter of formal state law.

Ironically, the protection of priorities as a way of guaranteeing new rights envisioned by the Reclamation Act of 1902 had unintended consequences for the future of the Rio Grande. Because prior appropriation was the only way to secure a water right under the system, many new rights were hurried into production. And the absolute preference for priority in time would tend to offer the law's most complete protection to rights that became more and more obsolete as the world around them changed.

Bien's code, with its three central elements, transformed these previous loosely enforced, local, or limited practices by formalizing them in a single body of law, raising that law to the state level, and centralizing water power there. The last step, which consolidated all of these elements, was the creation of the Office of the State Engineer. The idea was a new one. Wyoming had recently experimented with a professional water czar, but Bien's code innovated further by making the New Mexico state engineer the central state figure in the scientific and legal control of water. From his lofty position, the state engineer would apportion the waters of New Mexico among private beneficial users and permit new uses, all according to the new law and the best science. His position was a world removed from the traditional community acequia governments that depended on voluntary conformance to community values (motivated by avoidance of shame) and the exercise of personal responsibility to enforce a loosely defined water stewardship.

By 1905, only a year after he had proposed it, a proud Morris Bien could report that his draft code had been adopted "without material change" by the legislatures of North Dakota, South Dakota, and Oklahoma.[37] These states, unlike New Mexico, didn't have much of an existing tradition of water use and could easily lay down a fundamental water law because the foundation was so bare and smooth. The history of water use in New Mexico was much longer and the foundation considerably rougher, especially in the Rio Grande Basin, so the fit wasn't nearly as easy.

In 1905 the Territory of New Mexico adopted a very watered-down version of the Bien Code.[38] Reclamation officials were not pleased and said that they wouldn't recommend major federal projects in New Mexico if this was the best that the territory could do. Pushing hard for the federal money that would make projects like the Elephant Butte Dam possible, the Territorial Legislature caved in and, in 1907, adopted the Bien Code almost in its entirety, satisfying the federal government.[39] Science, law, and federal money would now flow to the Rio Grande where none had been present in large amounts before. The new forces were free to build tall dams to trap flood flows that were previously wasted down the main stem of the Rio Grande, confident that the underlying state law institutions were sufficiently leak-proof to guarantee the federal government the state-based water rights it was after.[40]

In a place as historically complex as New Mexico, there were bound to be some leaks in Bien's legal construction. Having just passed a law that repealed all previous laws, Bien's comprehensive code also adopted a separate statute that essentially exempted the existing community ditches in their internal operations from all of the code's formal requirements. The Rio Grande acequias were set apart from the law of the river and were allowed to govern themselves within their extremely limited domain.[41]

Where the two still intersected at the acequia diversions from the river itself, the courts were left to straighten out the relationship between the 1907 water code and the ancient acequia institutions over which it had been imposed. In Tularosa, a small town in south-central New Mexico, for example, a complex battle raged through the last decade of the nineteenth century and the first decades of the twentieth, pitting the town's community ditch against upstream developers who wanted to put the town's water to new uses. In 1914 the New Mexico Supreme Court ruled the ditch had no control over "unappropriated water" and owned no water rights. It ruled that only the new state engineer could define both the unappropriated water and the individual irrigators who owned the water rights delivered by the ditch. Isolated above and below from decisions like this, the acequias became antiquated institutions with little power.[42]

By the start of the twentieth century, the Rio Grande as a resource was under the control of the New Mexico state engineer and Reclamation. Together, in the name of development, science, and law, they proceeded to transform the Rio Grande.

Chapter VII

Big Dams, Irrigation Districts, and a Compact

Massive Engineering on the Rio Grande

On January 6, 1912, New Mexico was granted statehood, which occasioned both elation and frustration. The elation requires no explanation, but the frustration came from a sense that New Mexico's political stature was outstripping its economic progress. In particular, most of the population (as elsewhere in the United States at that time) lived by farming. In other parts of the United States, farmers were adopting the new scientific methods and machinery of the industrial era, but in New Mexico and also along the Rio Grande in northwest Texas, little had changed. Farmers were still diverting their irrigation water from an unregulated river and sending it down primitive acequias. Whenever snowfall in the San Juan Mountains was scant, the river dried up and crops failed. When runoff was plentiful, there were often destructive floods and seepage from the river, which caused the soil to become salinized.

Over the previous twenty years, the federal government had vacillated, first supporting a privately financed irrigation infrastructure, then opposing it, with no material progress. The problems only grew worse. This was about to change: over the next twenty years a major dam would be built to control the Rio Grande; two major, modern irrigation districts would be created; and flooding would be curtailed by massive levees. The increased ability to control the river led to increased concern that one party or another would divert more water than their entitlement. This mistrust threatened to bring development of the river to a halt. The issues were ultimately resolved only by negotiating an interstate compact. Although both the engineering infrastructure and the compact were initiated with specific goals in mind, the outcomes have often confounded those expectations.

A Dam Rises . . .

On June 13, 1913, the first concrete was poured for what would become the imposing structure of Elephant Butte Dam. On the Rio Grande, 124 miles north and upriver from El Paso, Texas, and 120 miles below Albuquerque, the U.S. Bureau of Reclamation was building a behemoth, the second-largest dam in the United States. Even today, the dam is the largest man-made structure sitting astride the Rio Grande. The dam was the centerpiece of Reclamation's Rio Grande Project, which included lands that extended along a narrow strip of the river 165 miles north and 40 miles southeast of El Paso. The purpose of the project was to provide a reliable and efficient source of irrigation water for southernmost New Mexico and the El Paso border area in addition to satisfying the international demands of Mexico. The project would ultimately include two major dams (Elephant Butte and Caballo), six diversion dams, 141 miles of canals, 462 miles of laterals, 457 miles of drains, and a hydroelectric plant.[1]

The second director of Reclamation, Arthur Powell Davis, had moved political heaven and earth to get Elephant Butte Dam authorized. And to build it, Reclamation literally had to move an extraordinary amount of earth. Before construction could commence, a camp for three thousand workers and

Men working on the Elephant Butte Dam construction site around 1910. New Mexico State (University Library, Archives and Special Collections) Image no. 02330090.

Elephant Butte Dam engineers admire their handiwork and the view through the spillway, May 1920. New Mexico State (University Library, Archives and Special Collections) Image no. 3240083. U.S. Bureau of Reclamation photo.

their families had to be built at the remote Elephant Butte site. The immense mass of concrete and rock required for construction forced Reclamation to lay thirteen miles of railroad track to the Santa Fe line at Engle, New Mexico. Then, in order to link the concrete mixing plant at the end of the railroad spur to the dam, the engineers hung a three-cable tramway system 350 feet above the river. The tram buckets that ran along its 1,400-foot length carried almost 2 cubic feet of concrete apiece and dumped an average of 250 cubic feet of cement every eight-hour shift into the dam below.

Over three years, Elephant Butte Dam slowly rose to its full height. Cement from high above rained down on hundred-foot-long block forms that were twenty feet deep. Large stabilizing boulders, called *plumbstones*, were carried on cranes, dumped into the partially filled forms, and embedded in the wet concrete. Teams of four or five men working below spaded the concrete against the forms and sides of the plumbstones. Once this process was completed for a section, a mixture of pulverized sandstone and Portland cement was sprayed on the upstream face of the gigantic dam. Some 611,000 cubic yards of it would be applied—shot from a 1¼-inch rubber hose—before the dam was finished in 1916.

Elephant Butte Dam, 1949. Photo by Len Sprouse, Courtesy Palace of the Governors Photo Archives (NMHM/DCA), no. 59169.

The crest of the completed dam spanned five hundred yards across the river. The base of the dam was one hundred yards thick, narrowing at the top to hold a road only sixteen feet wide. The pool behind it could store more than 2 million acre-feet of water spread over 35,000 acres. But it was the dam's height that measured its scale on the landscape. Rising to 306 feet high at its spillway, Elephant Butte Dam was junior only to the 349-foot Arrowrock Dam that had been completed on the Snake River in Idaho the previous year. These two tall dams dwarfed most of the many early federal dams in the West, including even the 280-foot Roosevelt Dam completed on the Salt River in Arizona in 1911. As wide at its base as it was tall at its spill-way, Elephant Butte Dam towered over the river it blocked.

And yet, the scale of the whole Rio Grande system still baffled man's efforts to control it. In places, the river made up in depth for what Elephant Butte Dam offered in height. Just north of Taos, New Mexico, the Rio Grande, over eons, had cut a six hundred-foot-deep gorge, twice as deep as the dam was tall, a compelling reminder of nature's ultimate control of the river.

. . . and Changes Everything Downstream

Mesilla Valley farmers in southern New Mexico had eagerly awaited completion of Elephant Butte Dam and Reclamation's network of new, modern canals and drains. The Mesilla Valley farmers had faced numerous periods when the river ran completely dry. In 1902, for example, the riverbed in the southern Mesilla Valley had been so dry it was used as a road.[2] The Rio Grande Project promised to resolve these problems, and in a technical sense the project was a great success. The vast amount of water stored in Elephant Butte Reservoir evened out the river flows, and the supply of irrigation water became reasonably steady and secure. The new canal system delivered the water efficiently. But, ironically, the new and improved infrastructure proved a greater threat to the farmers than had their old enemy, drought. By the mid-1920s, most of these farmers were gone.

The sudden abundance of irrigation water that the new dam provided proved irresistible to Mesilla Valley farmers downstream of the dam. They made up for years of deprivation by over-irrigating their lands with the now-plentiful water. Waterlogging very rapidly became a widespread problem, afflicting 70 percent of the farmland by 1917. Dam construction had reduced suspended sediments in the river flows. As a result, the clear irrigation water rapidly percolated into the ground, creating a need for more water to the fields, further compounding the waterlogging. Reclamation was consequently forced to construct an extensive network of open drains to correct the problem over the next eight years.[3] As the fields drained, alkalai deposits

A residence at San Elizario, Texas, in the El Paso Valley in 1919 shows the type of destruction alkali in the rising water table had caused by 1919. New Mexico State (University Library, Archives and Special Collections) Image no. 03240076. U.S. Bureau of Reclamation photo

remained in the upper soil layers, requiring substantial remediation efforts. The waterlogged lands had to be leveled, bordered, deeply plowed, and then subjected to heavy application of water, a process that cost approximately $35 per acre.[4]

Also, the Rio Grande Project improvements were not free. Better physical security came at the expense of economic security. Water that farmers had always diverted from the river at no cost except their own labor now had to be purchased. Taxes were imposed to pay for the costs of the huge dam and irrigation improvements. The repayment requirements of the project unleashed decades of protests and adjustments over the maximum costs of the improvements per acre and the period over which they would be paid. In any case, farmers who had been growing crops for their own subsistence, with only small surpluses, lacked a source of income to pay for water and land. Land values rocketed skyward: unimproved land that averaged $17.50 per acre in 1906 cost $50 to $60 per acre by 1913, with more developed tracts reaching up to $1,200 per acre by 1915.[5] The choice was clear for the original landowners: sell while the market was good or face crippling taxes on the increased value of the land. In some cases, land became delinquent because of nonpayment of taxes before the owners became aware of the situation and their own jeopardy, and someone else paid back taxes and received title to the property before the original owners could remedy the situation.[6]

The face of the land changed as well. Farms of increasing size filled in all available cultivable land in orderly geometric shapes. The amount of irrigated acreage in the lower Rio Grande valley more than tripled from pre-dam times to 1926.[7] A major change in the type of crop accompanied this shift: alfalfa, wheat, corn, and beans, which had been easily grown, made way for King Cotton, which was introduced to the area in 1918 and could thrive even on the moderately alkaline soils left behind from previous waterlogging. Within a couple of years, 90 percent of irrigators in the Mesilla Valley grew cotton, and by 1928 almost, 110,000 acres of irrigated Rio Grande Project land were cotton crops.[8]

With a secure water supply, the new commercial crops such as cotton were very lucrative, but the old-time farmers did not have the expertise to grow them successfully. To be profitable, such crops required large-scale mechanized farming, with extra manpower, draft animals, and equipment to clear and level fields, for which the traditional farmers did not have the necessary capital nor labor force. As a result of all these changes, the old families gave way to new blood, new ideas, and new property owners.[9] The number of farms in Doña Ana County, New Mexico, increased from 571 in 1900 to 1,054 in 1920, with most new farmers Anglo.[10]

Local farmers proudly display the first bale from the 1927 cotton crop of Doña Ana County, New Mexico. NM State Tourist Bureau, Courtesy Palace of the Governors Photo Archives (NMHM/DCA), neg. no. 005159.

Natural vegetation gave way to cultivated plants, with just a narrow belt of disconnected riparian forest lining the river by 1930. And because flooding continued after construction of Elephant Butte Dam, Caballo Dam was built two miles to the south, and the river was channelized between Caballo Narrows and El Paso. The bosque was removed from the former flood plain, and the river was confined to a narrow, deep channel placed symmetrically between setback levees. The river may have been tamed for farming, but residents came to realize the extent of what had been lost, and "especially the older Hispanic people spoke with fond remembrance of the walks they had taken in the bosque, the picnics and fiestas they had there, and how they used to go hunting and fishing along the river."[11]

The river water itself below Elephant Butte Dam was transformed by the more regular and clearer flows released from it. The dam had been built high for several reasons: so that its deep reservoir could accommodate the silt that engineers knew would accumulate, to minimize water loss to evaporation, and to provide the capacity for retaining sufficient water in wet years to get residents and farmers through the dry years. Yet the foresight of the engineers was not equal to the massiveness of the silt problem. The reservoir lost 16 percent of

Filling Up Elephant Butte Reservoir with Dirt

In 1917, shortly after the completion of Elephant Butte Dam, the government surveyed its reservoir floor.[1] The results caused consternation. The reservoir was expected to be usable for several hundred years, but the survey showed that of an initial capacity of 2.6 million acre-feet, sediment had already filled 50,000 acre-feet of the dam's capacity. At this rate, the reservoir would be filled with dirt by 2010!

As things turned out, in 2007 the reservoir still had a little more than 2 million acre-feet of capacity left.[2] What choked back the torrent of mud pouring into Elephant Butte?

Due to regular monitoring of the reservoir capacity by the Bureau of Reclamation and of the sediment content of the waters of the Rio Grande by the U.S. Geological Survey, the facts are easy to ascertain even if the causes are not. The rapid initial filling of the reservoir was part of the same outpouring of mud that buried San Marcial. But between 1915 and 1930, the rate of infilling dropped by a factor of two, to about 15,000 acre-feet per year.[3] It continued to drop, although more slowly, until about 1950, when it stabilized at approximately 3,000 acre-feet per year, an annual amount it has maintained ever since.[4] Sediment concentration measurements at San Marcial and upstream at Otowi tell the same story: infilling has slowed because sediment concentrations in the river and its tributaries have decreased, not because the amount of water carrying sediment has gone down.

The tributary that dumps by far the most sediment in the Rio Grande is the Rio Puerco. It is responsible for 30 to 80 percent of the sediment load of the Rio Grande at San Marcial, depending on the time period and the investigator.[5] The sediment concentration of the Rio Puerco is the third-highest of any river in the world.[6] However, this concentration also decreased by a factor of approximately three since officials began to measure Rio Puerco sediment loads in the late 1940s. The dramatic reduction has undoubtedly been a significant factor in slowing the rate at which Elephant Butte Reservoir is filling with sediment.

What was responsible for this region-wide outpouring of sediment and its subsequent slowing between 1920 and 1940? In the late 1800s, a network of arroyos began to slice through the tributary valleys of the Rio Grande.[7] During the first half of the twentieth century, these deep trenches began to heal, with vegetation becoming reestablished in the channels, flood velocities slowing, and more water soaking back into the channels, resulting in the sediment dropping out instead of being flushed into the Rio Grande.[8] But the ultimate cause of this cycle has remained somewhat controversial.

Increases in the frequency of high-intensity rainstorms and consequent natural evolution of the topography and geology of arroyo channels probably played a role.[9] However, the evidence also strongly suggests that removal of vegetation causes large increases in runoff and soil erosion.[10]

Many scientists of the early twentieth century linked the cutting of the arroyos to overgrazing by the large numbers of livestock that were introduced after the advent of the railroad.[11] The data subsequently seem to prove them right. Total livestock numbers in New Mexico declined by a factor of almost three between 1910 and 1940.[12] During the same interval, the sediment concentration of Rio Grande water at San Marcial went down by about the same amount.[13] During this time period, the Forest Service and other federal agencies began restricting grazing and logging and instituting erosion-control programs. The degree to which these measures were successful in containing erosion is difficult to quantify,[14] but a strong circumstantial case can be made that poor land-use practices in the Rio Grande highlands resulted in sediment inflows that obliterated San Marcial and threatened the lifespan of Elephant Butte Reservoir, and that improvements in those practices have brought the sediment supply down to a more manageable level. The Rio Grande sediment story shows once again that the consequences of upland land uses in a watershed collect in the rivers that lie at their heart.

Notes

1. Dortignac 1956, 40–45.
2. Ferrari 2008, 79.
3. Dortignac 1957, 43.
4. Ferrari 2008, 79.
5. Dortignac 1956, 49; Gellis 2006.
6. Gellis 2006.
7. Rich 1911; Bryan 1928.
8. Gellis 1992; Love 1997.
9. Bryan 1928.
10. B. Wilcox and Wood 1989.
11. Rich 1911; Calkins 1941.
12. Dortignac 1957, 57.
13. Gellis 1992, 81.
14. Gellis 2006.

its capacity to silt in its first 25 years, calling into question Reclamation's claims in the mid-1920s that it would last 233 years.[12] Previously, the high silt content of the unregulated Rio Grande flows had sealed the bottom of the river channel and helped to block water from leaking out into the lower, nearby fields. Now, downstream of the dam, the cleaner water scoured the river channel. The clear water seeped rapidly through the riverbed, reappearing as lakes and ponds scattered throughout the bottomlands. Elephant Butte Dam was clearly good for some things and people and bad for others.[13] In any case the dam had altered the river and the communities dependent on it.

Meanwhile, the Middle River Faces Its Own Problems

While Elephant Butte Dam was being constructed, farmers upstream in the middle Rio Grande valley appeared to be prosperous and secure in the centuries-long tradition of New Mexico's agrarian Hispanic settlers. Take, for example, Nicolas Sanchez, a farmer who resided in Albuquerque's North Valley. In 1914 Sanchez was fifty-seven years old and owned, free and clear, sixteen acres of irrigated land within the original Elena Gallegos land grant, less than a mile east of the Rio Grande.[14] This was sufficient for his family to farm and sustain itself, but was only a tiny portion of the original land grant named for Gallegos, a rancher who was the first woman in New Mexico with her own registered brand and who had acquired the full 35,084-acre grant in 1716.[15] For more than a hundred years the thousand acres north and south of the Sanchez property had been fed by the Gallegos community ditch and provided subsistence for him and his neighbors.[16] But the appearance of stability was an illusion; the buildup of salt on the land and episodic devastating flooding that had begun in the 1880s put middle Rio Grande farmers on an unsustainable path.

The larger agricultural community on which Sanchez depended had fallen into disarray. A particularly severe 1905 flood had ripped down an ancient river channel and cut right through the irrigated lands of Sanchez and his neighbors on the Gallegos grant. The flood had carried away the Gallegos community ditch headgate at the river and large parts of the ditch on which all had depended for water supply.[17] For many of Sanchez's neighbors, the massive flood was the last straw. They abandoned two compact, densely settled plazas near the ditch, including the one just south of the Sanchez farm, and moved to higher ground a few miles east. A few brave souls like Sanchez stuck it out, but without a supporting community, who could maintain the necessary ditches and headgates, even if the resources were available for the job?

To make matters worse, in 1914 Bernalillo County began the serious and systematic assessment of ad valorem property taxes, which were based on the value of the property. The burden on Sanchez was modest, but any tax levy at all was bound to strain a farmer who operated almost entirely outside of a cash economy.

Changes in the Sanchez land also made farming more difficult. Salt began to crystallize on the surface of his and his neighbors' land, poisoning traditional crops. Then, starting at the western edge of his tract, nearest to the river, his land first turned boggy, then grew a tiny pond in which the water level gradually rose and the shore expanded slowly eastward, consuming more and more of his land. By 1916, six of his sixteen acres of farmland were under salty water, and it looked as if the lake would continue to swallow more of it.[18]

Sanchez's plight was shared by most farmers along the middle Rio Grande Valley in the years before 1920. Growing costs, big floods, high water tables, and increased salt all threatened one of the oldest irrigation-based societies anywhere in the United States. The numbers told the regional tale. In 1922, there were about 40,000 acres irrigated in the middle Rio Grande, but another 40,000 acres were waterlogged and unusable.[19] What were Sanchez and his fellow farmers to do? There was talk about forming a flood-control and drainage district, but Sanchez's community viewed these proposals with more trepidation than anticipation. Would the old acequia associations control the new canals, or would outsiders? And how would Sanchez and his friends pay the taxes required to fund such a large engineering project?

Rise of the Middle Rio Grande Conservancy District

The United States wasn't about to come to the rescue of Sanchez or anyone else in the middle Rio Grande. The federal government had put as many resources as it could muster into the Elephant Butte Irrigation District and Elephant Butte Dam on the lower Rio Grande, as well as other New Mexico projects on the Pecos River. It couldn't—and wouldn't—take on more headaches in the middle Rio Grande. So the new state of New Mexico, barely ten years old, was forced to find its own solutions by means of a new Middle Rio Grande Conservancy District. The people with the money and vision to seek these solutions were for the most part Progressive Albuquerque businessmen and lawyers, not the small farmers. The primary concerns of the businessmen were property values, protecting their factories and stores from the flooding that episodically devastated them, and inducing new growth for the city. Their motivation for the proposed district was enthusiastically expressed in a 1928 editorial letter to the *Albuquerque Journal*: "[The district]

means that the value of suburban additions will be increased fully 800 percent, and that every irrigable farm in the valley will be increased from two to
500 percent in value."[20]

Although the irrigation district boosters were principally concerned
mostly with property protection and values, they also saw a need for great
improvement in agriculture. In their eyes, the criticism of W. W. Davis, a
mid-nineteenth-century immigrant, was just as valid in the early twentieth
century: "No branch of industry in New Mexico has been more neglected
than that of agriculture, which seems to be in about the same condition as
when the Spaniards first settled the country. It has been pursued merely as
a means of living, and no effort has been made to add science to culture in
the introduction of an improved mode of husbandry."[21] In the spirit of the
Progressive movement's emphasis on the efficient and rational use of natural
resources, the irrigation district promoters sought a solution that would lead
to better use of land and water to raise the general level of economic prosperity.[22] They hoped at one stroke to solve both the urban flooding and the
agricultural water supply and drainage problems.

Pearce Rodey, a leading New Mexico lawyer who was the driving force
behind the new MRGCD, looked to the cities of Pueblo, Colorado, and
Dayton, Ohio, for models of the kind of self-generated, locally financed
improvement districts that would allow the middle Rio Grande to save itself.
Reflecting his principal concerns, these cities had focused on flood control
in industrial settings, not irrigation reclamation. In 1923, Rodey went to the
New Mexico State Legislature and convinced it to authorize local groups to
create an Ohio-like conservancy district in New Mexico that could finance
itself by taxing local farmers and businessmen for the benefits that they
would receive from the district improvements.

Rodey's plan was greeted enthusiastically at first, at least in Albuquerque.
The process of initiating the district required approval of only one hundred
Albuquerque property owners, and Rodey had no trouble rounding up these
core sponsors. Anticipating protection from frequent devastating floods
at a relatively low cost, the Albuquerque Chamber of Commerce cheered
the proposal. So did the farmers engaged in large-scale market farming.[23]
But the plan had been conceived in collaboration with Albuquerque's business elite, politicians, and engineers, all mostly wealthy Anglos. This group
naively assumed that the small-scale Hispanic farmers who made up the
vast majority of rural residents would share their enthusiasm. In this, they
missed the mark.

Two issues spurred resistance to the district by the Hispanic villagers.
First, the benefits and the costs seemed unbalanced and unfair.[24] The project

was to be paid for by a tax on the amount that land values increased under the district's improvements. Because the bulk of the land was in the hands of small farmers who historically had been unable to afford land improvements, the estimated value of the improvements would be high. However, the farmers' cash flow was also so low that even a small tax increase would very likely cripple them. Meanwhile, wealthy businessmen in Albuquerque would receive invaluable flood protection for their large investments in facilities and goods in return for what were relatively modest tax increases. A second issue—local control—was almost as important to the Hispanic small-scale farmers as the increased tax burden.[25] The Hispanic farmers shared neither the businessmen's thirst for commercial growth nor the Progressives' passion for increased agricultural efficiency. For these farmers, the acequia was the very lifeblood of the community. They saw that the new conservancy district would replace their beloved community acequias with an engineered ditch under the control of unsympathetic outsiders.

In response to this perceived threat, a firestorm of opposition soon erupted among the Hispanic subsistence farmers north and south of Albuquerque. Before long, more than three thousand of them had signed a petition protesting the district's formation on the grounds that they had not been consulted and that the district tax assessments would ruin their subsistence farming. Piled on top of other property taxes that had only recently been enacted, the new district's special assessments for water improvements threatened to drown them in debt. The first of many law suits ensued. The New Mexico Supreme Court in 1925 upheld the constitutionality of the New Mexico Conservancy Act, which in 1923 had authorized the MRGCD, while noting in its opinion the massive opposition to it.[26]

Occasionally the opposition erupted into public confrontation. On April 19, 1929, only one month after actual construction of the district improvements had begun, a dragline operator sliced through the Los Chavez, New Mexico, acequia in order to excavate a culvert. A crowd of men from Los Chavez village, reinforced by sympathizers from Albuquerque, quickly gathered to try to save their water supply, and thus their crops. They insisted that the MRGCD officials who rushed to the scene guarantee the reconstruction of the acequia before they would let them leave. In retaliation, the leaders of the opposition were arrested when they returned to Albuquerque. The judge charged them with "a serious and sinister exhibition of ugly mobbery, which will not again be tolerated in this valley," even though no evidence of weapons or violent intent was ever presented.[27]

Learning a lesson from this vehement and unexpected resistance, the MRGCD quickly made significant concessions to the old farming communities it would now serve.[28] The district recognized existing rights to 80,000 acres of irrigated land: 40,000 acres that were still under irrigation and another 40,000 acres once irrigated that had become waterlogged and salinized.[29] The MRGCD contracted with most of the dilapidated community ditches to take over ditch maintenance and deliver water to the lands that they served.[30] In some instances, it curried favor by paying local Hispanics for the maintenance work that had historically been done for free out of local communal obligations.[31]

In order for the district to provide water management throughout the entire middle Rio Grande, it had to manage water on the lands of the six pueblos within the district. This required the district to obtain permission from the federal government. This federal permission was secured by the district by recognizing that the pueblos had 8,500 irrigated acres with rights superior to anyone else's in the district and by offering the pueblos the benefits of the improvements.

All of these proposed features were embodied in the MRGCD's plan, which was approved by the Conservancy Court in 1928. Called "Burkholder's Bible" after the district's dynamic first chief engineer, Joseph Burkholder, the lengthy plan consisted of three volumes. The level of detail was such that the first volume alone contained 435 numbered paragraphs, two long appendices, sixty-four tables, and too many maps to count.[32] The plan called for flood control, irrigation work, and drainage improvements on a massive scale.

Burkholder estimated in 1928 that the whole restoration project would involve repairing or constructing 500 miles of main irrigation canals and laterals, 400 miles of deep drains, and 190 miles of dikes and levees.[33] The district would move 25 million cubic yards of dirt and cost $10.5 million, a staggering sum in booming 1928, and one that would become simply unfeasible in the impending Great Depression. The United States agreed to pay $1.5 million of this amount as the pueblos' share. The $9 million balance would come out of money raised by district assessments on the 125,000 benefited acres.

For flood control, the plan did not call for the usual main-stem dam. Instead, it proposed a massive system of riverside levees from Cochiti to Elephant Butte to protect property near the river. The actual river channel would be narrowed and deepened so that the Rio Grande flow would stay put rather than meander through its broader flood plain—land that included a lot of valuable farm property. This improved channel would also increase the speed of river flows, promoting sediment transport and preventing aggradation of the riverbed.

The MRGCD would consolidate the seventy-two unique diversion dams of the community irrigation districts into four huge district diversions.

The community diversion dams had been crude, inefficient, and unable to control the flow of the river. The new diversions would be massive concrete structures reaching across the entire new channel of the river, able to control its full flow. Additionally, the MRGCD plan selected a site high on the Chama River at El Vado, New Mexico, for a dam to store supplemental water for middle Rio Grande irrigators.

Finally, the MRGCD would construct a system of drains as intricate and interrelated as the system of ditches had ever been. The drains would empty the lakes and lower the water table that had plagued area farmers like Nicolas Sanchez. As important, the salvaged water would be the property of the new district under existing New Mexico law. That new water would serve the 100,000 acres that the district proposed to restore to farming or to irrigate for the first time. Huge dredges and draglines appeared, gouging new fifteen-foot-deep, twenty-foot-wide drains in the centuries-old irrigated lands of the Middle Rio Grande, in many cases directly across the course of the ancient acequias.

In spite of the impending improvements, Sanchez evidently decided his situation was hopeless. He gave up his ancestral fields and left the land, most likely to seek hourly work in Albuquerque. When district inspectors showed up in the North Valley in late 1928 to assess the benefits that the project would bring to the sixteen-acre tract, they found a new owner, Pedro Armenta, in possession of the eastern eight acres that had been classified in 1917 by the state engineer as Cultivated Class I, the best condition possible. By 1928, only four of those acres were still being irrigated; the other four yielded just salt grass, a weed of almost no nutritional value even to livestock. A two-room adobe building in poor condition represented the only improvement on the tract. And yet, the MRGCD improvements would almost double the value of the tract, leaving Armenta responsible for paying for the increase.[34]

As for the eight eastern acres of the original Sanchez farm, they were now inundated by a swampy pond and belonged to the Sierra Meat and Supply Company. Even so, district inspectors found that the lay of the land and what they termed its "inherent fertility"[35] were average and that restoring the land to agricultural conditions through the proposed improvements would benefit the tract eightfold. The property owner would be required to pay for that benefit. The promised improvement was substantial, but the effect would be to add a significant dollar cost to a parcel whose highest and best use had once been subsistence and largely cash-free agriculture.

The MRGCD did fulfill its part of the bargain, however. In the North Valley, it hooked up the old Gallegos Community Ditch to a modern dam on the river, the Angostura Diversion. It rebuilt the ditch along its course and greatly improved its ability to convey water to the Armenta and Sierra Meat North Valley tracts. The MRGCD brought in huge dredges and drag lines and dug a massive drain twenty feet deep and thirty feet across, right through the lake that had swamped the tract and on through the rest of the waterlogged North Valley above and below. The salty water slowly percolated into the giant drain, leaving the soil once again fertile.

By 1935, when the first set of aerial photographs of the district was systematically assembled, anyone could see the improvements that the district had made in the North Valley and in general. The eight swamped Sierra Meat and Supply Company acres were now green with alfalfa, and the rest of Sanchez's 1916 farm was back in agricultural production.

And so it went, up and down the 125,000 acres of the Middle Rio Grande Conservancy District. Where ten years before, alkali wastelands had dotted the terrain and swamps and lakes and useless saltgrass ruled, vast acres of alfalfa and grains now appeared. Where ten years previously the river had wandered over, and often out of, a wide flood plain, it was now confined to a narrow channel. And where control of the water from the river had lain in the hands of community ditch commissioners and mayordomos, control now was vested in professional engineers, overseen, when necessary, by a judge relatively free of the vicious local politics.

Some of the farmers along the middle Rio Grande were delighted with these changes. These were mainly the Anglo newcomers or that minority of entrepreneurial Hispanic farmers who successfully adjusted to the new culture and technology. They produced the abundant crops of grain and alfalfa using modern, scientific methods of agriculture, fulfilling the agricultural vision of the Progressive reformers who had pushed the irrigation district.

This future had been foretold in three speeches given in 1930. The first was delivered at the formal initiation of MRGCD construction in March. Governor Richard Dillon proclaimed, "Conservancy brings the dawn of a new day!"[36] The day was indeed bright for the previously ruined farmland and the efficient farmers who now owned it.

The second, contrasting speech was not at the conservancy celebration, but was instead addressed to the Second Judicial District court by W. A. Sutherland, an attorney representing Hispanic farmers who sought greater representation in the MRGCD governance. He pleaded, "No other element except the agricultural has anything to fear from Conservancy. [City residents] have virtually nothing to lose if the project should turn out to be a complete failure. . . . They have everything to gain. But what about the farmer? He will lose all that he has."[37] And lose they did. Within twenty years, 41,000 of the 91,000 taxed pieces of land within the district would be delinquent in payment of taxes.[38] The MRGCD taxes were the last straw for many of the Hispanic subsistence farmers, who already were economically stressed by attempting to live off only the produce of their tiny plots of land. By the thousands they abandoned their ancestral villages and moved to the city.

The last of the three speeches was also delivered at the ceremonies celebrating the start of MRGCD construction and, ironically, it was the only one given in Spanish. "You cannot stop progress," B. C. Hernandez informed his Hispanic audience. "You may put obstacles in its way, but progress is bound to come."[39]

It is certainly unfair to blame the conservancy district for the demise of the traditional New Mexican lifestyle. Many of the Hispanic farmers had already abandoned their land before the district was established because their fields were ruined by waterlogging or floods. As elsewhere in New Mexico, new generations in local communities resisted the efforts of outside reformers to return them to an earlier way of life that was as harsh as it was admirable.[40] Even for those whose land was not damaged, life on the small farms in rural New Mexico was hard and the returns were meager. The children of Sanchez's generation, and those who followed, were gradually pulled away by the attractions of a wage economy and the excitement of life in bigger towns and cities.

Probably most of the movers and shakers behind the formation of the district genuinely thought that its benefits would improve the lot of the native farmers and help lift them into the twentieth century. It was the people of the land themselves who immediately perceived that the entrepreneurial, efficiency-driven vision from which the plans had sprung was inimical to their way of life and that the market economy required by the new MRGCD infrastructure would soon drive them from the land. They had been taught that hard physical labor, attending mass on Sunday, and carrying the santos in procession across the field would suffice to assure a secure future. Now they faced a world in which assurance and prosperity came through massive dams and dikes that bound the ancient Rio Grande to the will of man, and they did not have the wherewithal to pay their share of those costly works of engineering.

Prosperous investors such as Albert and Ruth Hanna McCormick Simms saw a great opportunity, and they quickly moved in to buy the formerly nearly worthless swampy fields. But everyone could not share their enthusiasm. For most of the indigenous Hispanic farmers, the MRGCD was an octopus that slowly squeezed them out of their livelihood. Just as was foreshadowed ten years earlier in the new, improved lower Rio Grande below Elephant Butte Dam, struggling farmers desperately switched cropping patterns in a futile effort to convert from subsistence farming to cash crops that

Elena Gallegos Land Grant: The Rest of the Story

No one looking closely could fail to see what changes the Middle Rio Grande Conservancy District had wrought in the human ecology of the valley where Nicolas Sanchez had first noticed a shallow lake mysteriously forming on his land. Where once existed a thriving, densely settled Hispanic farming community, there was now the immense estate of Albert Gallatin Simms and his pioneering feminist wife, Ruth. By 1940, the Simms's Los Poblanos Ranch had restored the land to agricultural production, but the ancient and traditional community that had farmed it for generations had vanished.

Simms could look out the windows of his Los Poblanos mansion and see fields of corn and wheat stretching off into the distance, framed by the far-off cliffs of the Sandia Mountains. The fields were all his.

Born into a prosperous though not wealthy family in Arkansas, he moved to New Mexico in 1912 and set up a law practice in Albuquerque in 1915. From that start, his career sped along from success to success. He became involved in banking, was elected to the city council, then the New Mexico House of Representatives, and finally to the U.S. Congress. When he arrived on Capitol Hill, he found that he was seated next to one of the very few women in Congress, the formidable and powerful Ruth Hanna McCormick. She represented Illinois and was a strong advocate of woman's suffrage. She was the first American woman whose face was featured on the cover of *Time* magazine (in 1928). Both were Republicans and widowers, and a romance soon blossomed between them. When their political careers ended with the Democratic takeover of Congress in 1931, they returned to New Mexico as a married couple. They set about laying plans for a residence that would reflect their successes in business and politics.

Simms began to accumulate land along the river bottom northwest of downtown Albuquerque. He purchased land from farmers stressed by district assessments; from state and district foreclosures that were forced on landowners who were unable to pay those assessments; and from a life insurance company that had picked up the entire Elena Gallegos land grant at a fire sale. By 1940, he managed to amass single ownership of a tract of land that extended from the Rio Grande on the west to the crest of the Sandia Mountains on the east. This included on its western edge the whole 300-acre irrigated tract of which Nicolas Sanchez's 16 acres had been a small part.

Albert and Ruth selected the most scenic location on the huge property and commissioned John Gaw Meem IV, the reigning architect of his day in northern New Mexico, to design a suitable house for them. The mansion, named Los Poblanos after the original settlers of the Gallegos land grant who came from the Mexican state of Pueblo, still stands today, surrounded by extensive organic gardens. Los Poblanos is a building of enduring beauty, with sweeping horizontal lines and understated Spanish themes.

The palatial aesthetic of Los Poblanos provides a forceful contrast to the tiny two-room adobe house of the Sanchez family that once stood on the land. Under U.S. rule established in 1846, new inhabitants had swept in, fired up by the opportunity to achieve economic success. Albert and Ruth Simms embodied the fulfillment of that vision. They used much of the money garnered during their ambitious careers to assemble the property that made up Los Poblanos from a hodgepodge of deteriorating tracts, return the land to verdant productivity, and construct in the midst of it what is perhaps the most attractive residence ever built in Albuquerque.

It was not the Simmses' fault that the impacts of runaway grazing and logging ultimately waterlogged Sanchez's little fields, or that assessments to pay for drains were beyond his means and those of other small farmers like him. Nevertheless, the conversion of the land that once sustained a close-knit agricultural community of more than one hundred people into the baronial residence of a single couple is emblematic of the sea change in ruling values that prevailed along the Rio Grande.

A formal dinner is served on the portico of Los Poblanos in 2009. Photo by Lois J. Phillips.

would support the new assessments for district improvements. And, as in the lower Rio Grande, much of the land quickly changed hands: land values escalated, Anglos replaced Hispanics, and large consolidated tracts replaced family plots.

Those new dams, dikes, canals, and drains also gripped the natural riparian system in an iron vise. Overbank flooding virtually ceased, and the world's longest riverine bosque running the 120-mile length of the middle river felt a shudder. A single, deep channel replaced the shallow, constantly shifting, braided channels of the Rio Grande, making it difficult for certain aquatic species to maintain a finhold under the new conditions. In the age of efficiency in which the district was born, no one much noticed these profound alterations to a functioning natural system.[41]

Who Owns the Flow? The One and Only Rio Grande Compact

While Albuquerque promoters cheered developments in the middle Rio Grande valley, lower Rio Grande interests in both New Mexico and Texas looked on with increasing alarm. From their downstream perspective, the improvements promised by the MRGCD's massive 1928 plan, including the proposed storage dam on the tributary Chama River at El Vado, directly threatened to reduce the supply of water into their $10 million Elephant Butte Reservoir. The MRGCD had similar concerns looking upstream: further expansion of the population and agriculture in Colorado's San Luis Valley would also undermine the district's water supply and $10 million investment. As for the Coloradans, the United States had denied them any increased Rio Grande storage since 1891. How did the New Mexicans of the middle Rio Grande think that they could build new storage when upstream Colorado couldn't?

The suspicions raised by the competing interstate claims of Texas, New Mexico, and Colorado to the shared Rio Grande had long simmered. The state farthest upstream, Colorado, produced most of the Rio Grande water in its highlands. New Mexico, the state in the middle, was squeezed from above by Colorado's control of the water sources and from below by Texas's claim to a share of the river. Poor Texas, so far downstream from the headwaters, had to worry about both Colorado and New Mexico. The new state and federal investments in infrastructure in the first quarter of the twentieth century brought these conflicting interests to a boil. A compact for the river was clearly needed, something along the lines of the Colorado River

Compact that had been signed in 1922 among seven western states, including New Mexico and Colorado.

Negotiations for such a compact began in earnest in 1928. To their dismay, the hydrologic engineers at the negotiating table found there was almost no data upon which to base the important discussions that preceded the proposed Rio Grande Compact. So they stalled; they made a temporary agreement in 1929 based on very general principles. This preliminary compact among the states guaranteed that while the needed data were being collected none of the three states could do anything to alter the 1929 status quo on the river. For the next ten years, bitterness escalated, offers and counteroffers proliferated, and the technical data deepened, but one thing held steady: the final 1938 Rio Grande Compact froze interstate rights and obligations at the level of the 1929 conditions on the Rio Grande.

The journey from the 1929 agreement to the 1938 compact traversed a maze of high-stakes law suits, engineering, and politics that would have puzzled the most astute citizen. But the need for a permanent agreement became increasingly apparent. Interstate claims had legal primacy and had to be cleared up before any state's plans for the future could be secure. In 1938, in a dispute between Colorado and New Mexico over the puny La Plata River, the U.S. Supreme Court ruled that the shared interstate rights to a water source came before the rights of any individual state to any water under its own law.[42] The three states had each secured substantial rights to the river under their own state laws: Colorado, for the San Luis Valley agricultural and transportation interests; New Mexico, for the MRGCD; and New Mexico and Texas, for Elephant Butte Dam. But the real security of that water depended, as it turned out, on the relative rights of Colorado, New Mexico, and Texas to the Rio Grande as a whole. They would have to straighten out their relative rights to the river before any state could feel secure in its own individual allocation.

Nevertheless, the compact negotiations dragged on at a snail's pace. Faced with the physical reality of the just-completed El Vado storage dam on the Rio Chama, which was beginning to store Rio Grande water otherwise bound for Elephant Butte Reservoir, Texas in 1935 became so frustrated that it momentarily abandoned negotiations. It sued New Mexico in the U.S. Supreme Court to have the Court equitably apportion the river, because the states apparently couldn't do the job on their own.

The Supreme Court's special master (a person appointed by the Court to help in water adjudications) held thirty-nine hearings in Albuquerque and El Paso between 1935 and 1937, producing 3,000 pages of conflicting testimony by technical experts and 267 exhibits of all types. In March 1937 the special master postponed the proceedings, pending one last shot at negotiating a new and permanent compact. As he did so, a torrent of new data generated by the federal Rio Grande Joint Investigation, a massive project to understand the river—one that cut across a host of federal agencies and scientific disciplines—was released.

Texas's lawyer in the 1935 Supreme Court lawsuit was a young Las Cruces, New Mexico, attorney, Edwin L. Mechem, who went on to become a three-time governor of New Mexico, a U.S. senator, and a bulwark of New Mexico water interests. The fact that Mechem got his start representing Texas, New Mexico's most bitter rival, provides a hint of the complexity of the alliances and shifting interests that lay beneath the surface of the final Rio Grande Compact negotiations.

Those final negotiations also introduced perhaps the most critical players, the part-engineer, part-lawyer enginoirs. As data poured in from the court documents and the Rio Grande Joint Investigation, compact negotiators depended more and more on the esoteric analyses of these advisers. In the end, the 1938 compact was really the enginoirs' product, and the vision of the river on which it was built was an engineered river.

Most of the engineering advisers were seeking surface flow data that actually reflected 1929 Rio Grande water use in the three states. Once it was known what Colorado and New Mexico used, Texas (and its southern New Mexico partners) would get what was left over. As it turned out, the upstream claims of Colorado and the downstream claims of Texas were easy to deal with. It was the middle-reach claims of New Mexico within the newly constituted MRGCD that proved especially troublesome.

A Static Compact on a Dynamic River

On paper, it's not hard to allocate streamflow to meet numerous demands, but to make it actually work on the ground and in the river requires vast amounts of know-how and even arrogance. The sources for streamflow in the middle Rio Grande varied greatly along the length of the river. The upper end (represented by the Otowi, New Mexico, gauge) depended on snowmelt from the mountains north of the region, but the downstream flows were strongly affected by the highly variable summer monsoon rain that fell south of Otowi and north of Elephant Butte. To complicate matters, New Mexico's depletions from the river under the baseline 1929 conditions took a higher percentage of the available supply during years of low flow than they did in years of higher flow. How could the engineering advisers possibly construct a schedule of deliveries along the length of a complex river that would be both sufficiently understandable for practical use and perceived as fair enough to prevent further interstate disputes?

A leading force among the enginoirs was the engineering adviser for Texas, Raymond A. Hill, a consulting engineer from San Francisco. Already experienced in water matters across the West when he was hired by Texas in 1935, Hill never lacked self-confidence. He felt that he alone could make sense of the technical data flooding the compact negotiators. So sure was he after the 1938 compact negotiations of what the negotiators had done that he took it upon himself to set down in writing what the Rio Grande Compact meant.[43] So self-confident was he during the final negotiations that observers conferred on him superhuman, if not entirely complimentary, characteristics. "Raymond Hill," they said, "was the only engineering advisor who could strut sitting down."[44]

In a series of intense meetings held in late 1937 and early 1938, the engineering advisers, led by the strutting Hill, did come up with a workable schedule that would be embodied in the 1938 Rio Grande Compact. To address the vicissitudes of winter snowpack versus summer torrential rains as the source of middle Rio Grande water, a ten-year period was set before compact violations could begin to allow the accumulation of debits and credits for each state. Additionally, the engineers adjusted New Mexico's entitlement with respect to the erratic middle Rio Grande flows: the entitlement would decrease as flows past the upstream Otowi gauge increased, to the point that New Mexico would be required to pass along to Elephant Butte Reservoir 100 percent of flows at Otowi in excess of 405,000 acre-feet per year.

With these practical strategies added to the compact, the three states finally had what they had wanted for so long: a firm, measurable definition of the 1929 condition that would both limit and entitle each state to its share of the river. On that firm basis, the commissioners representing Texas, New Mexico, and Colorado signed the Rio Grande Compact in 1938, and the following year the respective state legislatures and the U.S. Congress ratified the agreement.

For the foreseeable future, the water supply that each state would receive from the Rio Grande would be fixed and limited by the amount of water passing one gauge and reaching the next. (To that extent, the 1938 compact harkened directly back to the Rio Grande gauging that the United States began at Embudo, New Mexico, in 1879.) Under the new compact New Mexico would be responsible for everything that happened to the river between the Otowi gauge and Elephant Butte: what the state did, what cities and private parties did, and what nature did. And yet Hill and the other engineers were so confident of technology and science that they fully expected New Mexico could manipulate anything between the two gauges—nature, the river, the water uses—and so keep New Mexico in compliance without sacrificing control over the river.

The compact became the fixed law of the river. As it became New Mexico's fundamental law, causality began to flow in a new direction. The compact was originally designed to match the requirements specified by the hydrological conditions on the Rio Grande, but soon the Rio Grande would begin to be redesigned to meet the requirements of the compact.

Chapter VIII

Mount Reynolds on the Middle Rio Grande

In 1938, New Mexico began to both enjoy the assurances of the Rio Grande Compact and incur its obligations. The assurance was the promise of an equitable commitment of water flowing downstream to New Mexico from Colorado, and the obligation was to supply a similarly equitable amount over Elephant Butte Dam to the Mesilla Valley and west Texas. The security of the upstream supply proved to be somewhat illusory. More critically, over the next seventeen years, the state found itself falling farther and farther behind on its downstream obligation, to the extent that the federal government threatened to take over the system on behalf of downstream farmers.

Into this mess in 1955 stepped a relatively unknown mechanical engineer and academic researcher named Steve E. Reynolds, to assume the post of New Mexico state engineer. Although his qualifications appeared meager, he came equipped with a strong fundamental understanding of the hydrology of the Rio Grande, which he used to modify both the regulations governing the use of water and the river itself. He slowly began to pay down New Mexico's by then enormous water debt. Riding on his successes, he would continue in office for an unprecedented thirty-five years. Reynolds's legal and administrative innovations produced a permanent shift in the water law of much of the United States, while his engineering solutions mutated the Rio Grande from its ancient form. Probably more than any other person, Reynolds is emblematic of the human forces that reined in the Rio Grande. His legacy is that he made New Mexico face hydrologic reality even while exploiting that reality to maximize an ultimately unsustainable beneficial use, and he used technology as much as possible to make nature yield to the compact and its legal requirements.

New Mexico Falls into Debt

For the first three years after the Rio Grande Compact was signed—from 1938 to 1941—New Mexico was able to send surplus water past Elephant Butte Dam. But in 1943, it began to miss its assigned delivery. After that, through wet years and dry years, New Mexico's water debt mysteriously continued to grow. The compact, which was the law of the river, clearly stated that "the accrued debit shall not exceed 200,000 acre-feet at any time," but by 1951, New Mexico passed that limit. By 1955, the water debt mounted to over 450,000 acre-feet.[1] Texan and New Mexican farmers below Elephant Butte Dam who had thought that the compact would solve their water-supply issues were now furious over being denied the water they were legally owed and needed desperately for their crops.

Beginning in 1950, drought deepened the crisis, reaching down from the Midwest and Great Plains to New Mexico by 1954 and encompassing a ten-state area that included the Midwest, the Great Plains, and New Mexico. Although this regional drought eased in 1957, from 1956 to 1965 the Rio Grande produced good runoff for only three of the ten years, and New Mexico's water debt grew ever faster. The compact had stipulated that New Mexico would owe an increasingly smaller proportion of the flow measured at the Otowi gauge in northern New Mexico when runoff was low, but it failed to deliver even that.[2]

Enter Steve Reynolds

So when Reynolds was appointed to the prestigious post of state engineer in late summer 1955, he was in trouble from the start: from the drought, from the growing debt of water deliveries past Elephant Butte Dam to farmers in southern New Mexico and Texas, and from legal threats.

In fall 1956, a year after his appointment, Reynolds went to New Mexico State University in Las Cruces to deliver his first state-of-the-water address to the annual New Mexico Water Conference. Reynolds used the opportunity to summarize his first year on the job and describe the water situation across New Mexico. He ended with a characteristic self-deprecating joke, the first of many in his thirty-five year tenure as "New Mexico's water boss," as the *Wall Street Journal* later called the position.[3] Reynolds said that he "found himself dumped into this bewildering maze of water resource problems about a year ago." To explain his predicament he drew on a Rudyard Kipling poem he had learned as an Illinois school boy, the one that began and ended with the famous lines, "If you can keep your head when all about you are losing theirs

. . . Yours is the Earth and everything that's in it . . ." In Las Cruces, Reynolds changed the final line and ended his 1956 address with a new perspective on the Rio Grande state of affairs he had inherited: "Then you probably don't understand the situation."[4]

The son of an Illinois farmer active in the Midwest Socialist Party, Reynolds had come to New Mexico in 1935 to attend the University of New Mexico (UNM) and play football. On the day he arrived, the car in which he traveled got stuck in the sand of a dry arroyo. As Reynolds and his companions dug and pushed, a thunderstorm erupted and sent a wall of water down the arroyo, tossing the car and the Illinois emigrants topsy turvy. "What better introduction to New Mexico and the Rio Grande?" Reynolds used to ask.[5]

At UNM, Reynolds excelled as a student and athlete, graduating first in his 1939 class, an unheard-of accomplishment for a mechanical engineer, let alone a captain of the football team. While at UNM he befriended fellow student John Simms and fell in love with Jane Iden, the beautiful daughter of one of the state's preeminent attorneys. To cap these personal and political successes, Reynolds was elected president of his class.

After graduation, he returned to UNM as an assistant professor of engineering and an assistant football coach. In the latter role, Reynolds coached a young New Mexico player, Bruce King, who would go on to become a three-time governor of New Mexico. At the start of each of his terms, King reappointed Reynolds as state engineer. Reynolds enjoyed the reversal in roles, quipping in his self-deprecating style, "I'm the only guy in the world who started out as the governor's coach and ended up as his water boy."[6]

After World War II, Reynolds went to work for E. J. Workman, the irascible and dictatorial president of the New Mexico Institute of Mining and Technology (New Mexico Tech) in Socorro, seventy miles south of Albuquerque. Workman had a passion for the physics of thunderstorms. Reynolds was one of Workman's top researchers, running elaborate field experiments on cloud seeding and flying in the nose cone of a B-49 into thunderstorms above Santa Fe.[7] Then-governor John Simms plucked him from the ivory tower of atmospheric research and set him in the middle of the combative arena of New Mexico water politics. Why Simms picked Reynolds for the critical post of state engineer has never been entirely clear. Reynolds was a mechanical engineer and thunderstorm physicist, not a hydraulic engineer nor a water law expert. However, it seems likely that Simms picked Reynolds because they had known each other at UNM; because Holm Bursum Jr., a Socorro political powerhouse, recommended him; and because Reynolds's boss, the respected Workman, probably backed him. In addition, Reynolds, although a relative newcomer to New Mexico, already had a cachet with powerful forces in state politics.[8] Asked later not how he got the job but how the job got him, Reynolds said that his UNM friend Governor John Simms had cajoled him into it. "I really didn't want to leave my research," he told a reporter, "but Johnny (Simms) delivered me a lecture on citizenship and responsibility. I figured that I'd serve through his administration, but then I just never found a good quitting place. There was always something that I wanted to finish."[9]

Reynolds had a certain aura—the certainty and charisma of a winner that must have been enough in Simms's eyes to compensate for his lack of traditional water credentials. Reynolds was a very bright man, but many in the water establishment of 1950s New Mexico must have thought that he was wet behind the ears and poorly qualified for such a critical position.

Reynolds brought with him a faith in public control of public resources that he had learned from his socialist father. He also harbored a healthy skepticism about pie-in-the sky hopes for increasing New Mexico's water supply by seeding clouds or desalinating vast underground pools of brine. As a scientist, he recognized that nature and Rio Grande water obeyed certain physical rules and natural laws. As an engineer, he held a strong belief that humans had plenty of room to manipulate the environment—within the constraints imposed by these natural laws—to further their own interests. Reynolds brought these tenets to bear on the complex Rio Grande situation that he had inherited. In 1955 he started to set the Rio Grande on his own course, for better or worse, for the next fifty years.

Facing the Rio Grande Crisis

The Rio Grande Compact had assured downstream irrigators that they would get an equitable share of Rio Grande water even under conditions of drought. When water deliveries failed to arrive, furious Texas farmers turned to their only source of redress under the interstate compact: the U.S. Supreme Court. In 1951, Texas sued New Mexico and Colorado for underdelivering Rio Grande water required by the compact. Texas also complained that New Mexico continued to store water at the Middle Rio Grande Conservancy District's El Vado Dam even when not enough was getting to Elephant Butte Dam, in clear violation of the compact, and asked the Supreme Court to take control of the Rio Grande and straighten out the interstate mess.[10] The suit embodied the worst fears of New Mexico and every other western state: that a federal administrator would take over the river. A sword of Damocles hung over Reynolds's head as soon as he arrived in Santa Fe to take his new position.

At the same time that Texas's complaints about the middle Rio Grande were ripening into a lawsuit, the MRGCD was falling apart. Envisioned in the 1920s as the savior of the river and the communities dependent on it, the district, by the late 1940s, was being swamped by the same problems that had plagued the middle Rio Grande only twenty years before. The district's ditches were in disrepair, again. Its drains were blocked with weeds, again. Its flood protection had not been able to withstand big pulses of water on the main stem. Its jurisdiction was not broad enough to halt the damaging grazing and logging practices that poured sediment into the river. And it was broke, broke, broke.[11]

The MRGCD was broke, most fundamentally, because it was launched at the beginning of the worst economic depression in living memory. It was broke because it couldn't repay the bonds it had already sold to outside investors to fund the project. Under those circumstances, no one was willing to invest in new MRGCD bonds to produce more revenue. And it was broke because many of the landowners who had received the benefits of district improvements had been unable to pay the assessments on their land. By 1949, 41,000 of the 91,000 taxed pieces of land within the district were delinquent, and the state of New Mexico ended up technically owning most of the land. The MRGCD itself had no operating funds.[12]

New Mexicans turned to a common practice in times of local financial trouble: they asked the federal government to bail them out. And so the Bureau of Reclamation, which had brought Elephant Butte Dam to the lower Rio Grande, turned its attention to the middle Rio Grande after the end of World War II. The bureau considered straightening out and deepening the river between Espanola, New Mexico, and Elephant Butte so that the channel could better transport its pulsing flood flows. It considered digging a channel through the jungle of deep-rooted vegetation at the head of Elephant Butte Reservoir that prevented what little water made it that far downstream from actually getting into the reservoir.[13]

In the end, Reclamation agreed to take over the whole MRGCD and in 1951 acquired all of the district's assets—its ditches, diversion dams, and water rights—as well as its liabilities (primarily the outstanding bonds) in exchange for the promise of the federal government to spend $18 million to re-rehabilitate the river and the MRGCD system. Reclamation would operate the complex conservancy district system of diversion and distribution of Rio Grande water, while district users would pay back the $18 million over forty years and interest free.[14] By the time Reynolds arrived as state engineer in 1955, Reclamation's projects for the river and the district had made little progress, since the money to support them still had to be appropriated by Congress.

The U.S. Army Corps of Engineers was also getting ready to go to work on the middle Rio Grande. A 1947 memorandum of understanding between the two federal water agencies, Reclamation and the Corps, divided up Rio Grande water duties, with reclamation activities going to Reclamation and flood control to the Corps. Congress confirmed this division in the 1948 Flood Control Act, the magna carta for federal work on the Rio Grande.[15]

In the early 1950s one last big piece in the Rio Grande puzzle was Albuquerque. Its population had exploded in size after World War II, shooting from 35,000 in 1940 to more than 150,000 in 1950, and 15,000 new residents arrived yearly. The frenzied growth was fueled by the belief that there was plenty of land for new homes, mostly on the mesas east of the river, and plenty of water. The city fathers and developers believed that Albuquerque sat atop a huge groundwater basin, perhaps the size of Lake Erie, offering an almost infinite pool of good-quality water in the middle of the desert. They saw no connection between their heavy groundwater pumping and low flows in the river, nor did they look for one. The Rio Grande Compact was a surface-water agreement that had nothing to do with groundwater. The abundant groundwater supply, situated as it was beneath the path of an interstate river, seemed to be an almost providential gift that enabled unlimited growth. The city fathers saw no reason to look a gift horse in the mouth.[16]

Tracking Down the Missing Compact Water

The most immediate crisis facing New Mexico and its new state engineer in 1955 was the state's thirteen-year-long failure to meet its compact-mandated water delivery obligation to the lower Rio Grande and Texas. What had gone wrong? The technical terms of the compact had been hammered out very carefully to ensure a fair water supply to everyone, under dry conditions and under wet conditions. Yet even in the relatively wet 1940s, New Mexico was far from meeting its water obligation. The continuing drought promised only to exacerbate the already disastrous situation and thus provoke the U.S. Supreme Court to take control of the Rio Grande out of New Mexico's hands. Unless the new state engineer could solve this problem that had baffled his predecessors, he would soon go down in ignominy.[17]

In 1955 the position of state engineer must have seemed like a swift way to end a promising career. Reynolds, however, had an advantage. Even before he was appointed, he knew at least part of the answer to the problem of making future deliveries under the compact. Through a remarkable coincidence, the two men who were likely the best qualified in the world to understand the groundwater portion of his delivery dilemma had moved to New Mexico, one of them almost next door to him. Together the two men showed State

Engineer Reynolds that pumping groundwater threatened the compact-agreed river flows.

Reynolds's Socorro neighbor in the 1950s was a young, severe, mathematically brilliant Iraqi expatriate named Mahdi Hantush, an early expert in the arcane science of well hydraulics.[18] During the time he worked at New Mexico Tech, a span that overlapped with Reynolds's employment there, Hantush published a commentary in which he improved the equations used for describing how wells near rivers get part of their water from the river and not just the aquifer alone.[19] Although this topic might have seemed interesting but unimportant to Reynolds the lightning researcher, it provided a critical insight for Reynolds the state engineer, who was now responsible for explaining the cause of New Mexico's increasing inability to deliver under the compact and ensuring future deliveries.

The second man who knew the answer to at least part of Reynolds's problem lived in Albuquerque. That man, C. V. Theis, was the district geologist of the Ground Water Branch of the U.S. Geological Survey. Theis had published in 1935 what is arguably the most cited paper in the history of groundwater hydrology, formidably titled "The Relation between the Lowering of the Piezometric Surface and the Rate and Duration of Discharge of a Well Using Ground-water Storage."[20] This paper provided the theoretical foundation for nearly all of Hantush's work, and indeed, for most of groundwater hydrology of the time. However, Theis was an intensely practical person, and he followed this theoretical paper with a much more applied paper in 1940, in which he pointed out that water pumped out of wells not only comes from the aquifer, but is also drawn from rivers or lakes that are within the area affected by the wells.[21] In 1941, he developed this insight into an equation that quantified the depletion of the streamflow.[22]

The combined insights and proximity of Hantush and Theis led Reynolds to an important realization: groundwater pumping would have to be controlled in order for New Mexico to meet its Rio Grande Compact obligations. The problem revolved around pumps. In the 1930s, when the compact was being negotiated, the only pumps available for wells were slow, clunky models actuated by pistons. But a new technology became available at the beginning of World War II: powerful downhole turbine pumps.[23] In a desert region where lakes, rivers, and even reservoirs are widely spaced and unreliable, the powerful new pumps were a godsend. Albuquerque saw them as a means to guarantee a water supply for a growing city. Businesses drilled many wells to get clean water for their industrial needs.

But the most enthusiastic well drillers were the farmers who could finally escape the restrictions imposed by the fickle flows of the Rio Grande. Most farmers still used irrigation water from the ditches when it was available, since it was effectively free and the turbine pumps consumed expensive electricity or gasoline. But when river flows diminished in late summer or even earlier if there was a drought, irrigators simply switched on the supplementary groundwater pumps and the problem was solved. At least this solution was to the farmers much less complicated than trying to use the legal basis of priority to shut down the City of Albuquerque municipal wells, which had junior rights.

Where did this new supply of water come from? Almost nobody worried about that relatively theoretical question. The new water was abundant and pure, and it gushed from the ground at the flick of a switch. Why question Providence?

Theis was one of the few people who had worried about the source of well water, and he had done so for years. In the late 1930s he had investigated why flow in the Pecos River of eastern New Mexico had diminished. The Pecos River valley had a very productive artesian aquifer and flowing wells were extensively used for irrigation. Theis began to suspect a connection between reduced river flows, lowered groundwater, and the flowing wells, a suspicion that he later proved mathematically.[24]

Theis next applied his new insight to the Rio Grande valley. He did not have any audience for this new work until 1953, when he was invited to give a talk to the Albuquerque Chamber of Commerce on groundwater and wells in the city. The members of the chamber probably expected a dry and technical summary of regional hydrogeology. What they actually got was a stern warning.

Theis began his lecture by lambasting the local leaders:

> Albuquerque [has] never expressed any interest in the status of its groundwater supply and . . . never asked a question about it. . . . Two or three years ago some wells quit pumping water. Responsible officials of the city said that the water table dropped 35 feet overnight and no one could have foreseen it! What happened was that the city got a notice from its bank that its account was overdrawn and when it complained that no one could have foreseen this, the city only said in effect that it had no bookkeeping system![25]

Theis then went on to explain lucidly the hydrological principles involved:

> All water that enters the aquifer is eventually discharged naturally through springs and seeps. No water in any significant quantity is added or subtracted underground. A well is a new discharge superimposed on

this previously stable system. As a consequence, all the water pumped by the well is balanced by (1) an increase of recharge, (2) a decrease in natural discharge, or (3) a reduction of storage in the aquifer—or by some combination of these. In the Albuquerque area we cannot increase the recharge without some artificial construction. *Hence all the water pumped by Albuquerque has been represented and all the water pumped will be represented either by less storage, meaning a fall in water level, or less water draining to the river.*[26]

Theis concluded his talk with this clear warning: "Properly located wells can continue to satisfy Albuquerque's needs for the foreseeable future, but it must not be overlooked that they are still a diversion of water from the Rio Grande."[27]

Reynolds, with his training in engineering and physics, was quick to absorb both the difficult mathematics of Theis's and Hantush's theoretical work and the very practical implication of their conclusions. The water that everybody pumped so happily from their new wells was not something for nothing. There was no free lunch, not even for the Albuquerque boosters at the Chamber of Commerce. At first nearly all the water pumped from a well came from storage in the aquifer, but as time went on, more and more of the water was being sucked out of the river. Theis's calculations showed that after only five years of pumping, a well located a mile from the river would pull 75 percent of its water straight from the Rio Grande.

The flow of the Rio Grande was theoretically protected from further surface water diversions by the Rio Grande Compact, but the ever-increasing number of new high-capacity wells eventually and inevitably would constitute exactly what Theis had called them: a new diversion of water from the Rio Grande. If the flow of the river wasn't then quietly being siphoned off under the very eyes of the bemused bureaucrats and lawyers who were charged with meeting the compact and protecting the users of the Rio Grande who had the oldest rights, it inevitably would be.[28] The amount of water disappearing due to these unauthorized "diversions" was only a part of New Mexico's water deficit in the early 1950s, but Reynolds realized that pumping was bound to increase along with the growing population. New Mexico was doomed to permanent water bankruptcy under the compact unless the well drillers and pumpers could be brought under control.

Reynolds Gets Groundwater Under Control

Reynolds had the answer to at least part of the mystery of the disappearing Rio Grande water, but managing the river over the long haul meant that he had to make New Mexicans face hydrological reality. The fact was that in the Rio Grande basin, surface-water irrigation and groundwater wells drew from the same source. For a while, both irrigation and pumping could increase because of the long-delayed impacts of wells on the river, but when these depletions finally impacted the river, as they eventually would, either the surface-water use or groundwater use would have to cease in order to bring overall depletions back to the levels authorized by the compact.

To bring this scientific reality to the law of the Rio Grande, Reynolds had to ascertain and take control of both the amount and location of groundwater pumping throughout the developed part of the Rio Grande valley. This was a task guaranteed to put Reynolds on everybody's hit list. Farmers pumped groundwater, lots of it. Developers had to pump groundwater to supply their burgeoning new housing tracts. Industry pumped groundwater. The City of Albuquerque pumped vast amounts of groundwater. Reynolds would have to challenge all these powerful interests to wrest away what they considered their God-given right to pump as much groundwater as they wanted. The battle to bring groundwater under his control was one that pitted Reynolds against nearly every other interest in the Rio Grande, but it was a battle he had to win.

Less than one month after he addressed the annual water conference in Las Cruces, New Mexico, in 1956, Reynolds declared the middle Rio Grande from the Colorado border to Elephant Butte Dam a "groundwater basin" to be administered by him as state engineer. New Mexico had forged the unique new concept of groundwater basins in order to control withdrawals on the Pecos River in the 1930s. Now Reynolds extended state management of groundwater to the Rio Grande.[29] In effect, with this declaration Reynolds seized public control of the groundwater on both sides of the river.

As Reynolds undoubtedly anticipated, the farmers, private land developers, and their City of Albuquerque boosters kicked and screamed and finally sued. After a long and contentious legal journey, the case reached the New Mexico Supreme Court. The court justices looked at the scientific evidence for the connection between the Rio Grande and the aquifer and concluded that if the state engineer had the authority to ensure water deliveries under the compact, then he would have to be able to control pumping. When they were unsuccessful in court, the protesters pushed through the state legislature a law that nullified Reynolds's authority. However, Reynolds had at least one powerful ally, Governor Edwin L. Mechem, who backed Reynolds by vetoing the bill. Finally, by 1963, the New Mexico courts upheld Reynolds's authority and explicitly authorized him to manage the interrelated Rio Grande groundwater resources.[30] Reynolds's victory was momentous. From now on, water law had to contend with the hydrological reality that surface water and groundwater are interconnected and that exploiting one will affect the other.

From the moment he seized control over the groundwater that fed the Rio Grande, Reynolds had a clear plan to manage the resource. In order to link the legal and hydrological realities, he invented the concept of *offsets*. Putting into practice the concepts presented in Theis's 1953 argument to the Albuquerque Chamber of Commerce, Reynolds formally acknowledged that over a long period of time, taking water from wells in the Rio Grande basin would deplete the surface water flows of the river by the amount pumped from the wells. Due to his new understanding of groundwater hydraulics, Reynolds appreciated a crucial fact: the amount of water sucked from the river depended not only on the rate of pumping, but also its duration and the location of the well. Wells close to the river would start to deplete the river in as little as five years. But the impacts from wells miles away from the river might not be felt on the Rio Grande for as long as fifty years, though they would inevitably equal the total amount of water pumped. In that long meantime, groundwater could be pumped without further reducing river flows in violation of the compact and the prior existing rights of farmers.[31] Reynolds required groundwater pumpers to calculate their depletion of the flow of the Rio Grande using a scientific formula and then purchase (but not use) surface water rights to offset what would otherwise be a reduction in the flow of the river. This exercise effectively allowed pumpers to gradually purchase the necessary surface water rights, while phasing in the new legal constraints.[32]

Between the time when pumping began and when the full effect would be felt on the river, pumpers were able to obtain some groundwater without having to pay for a surface water right. Because of this, the inhabitants of the Rio Grande valley were temporarily able to use more water than their allotment under the compact. But even the amazing Reynolds could not beat the physical law of conservation of matter. He recognized that the day would come when the combined surface and groundwater uses in the Rio Grande valley would have to shrink back to their compact-authorized and limited levels. He foresaw a time, perhaps around 2006, when every drop the wells drew would be coming from the river: the groundwater pumped would no longer be a separate resource but would be purely groundwater that fed the river's base flow. Until then, New Mexico could keep pumping groundwater at a rate that was faster than it was recharged, as an unappropriated resource. Once the effects of pumping hit the river, however, compact obligations to Texas farmers would have to be met by stopping an equivalent amount of surface water diversions, such as to acequias and the MRGCD.[33] Reynolds's complicated scheme for continuing to pump groundwater while still protecting senior surface water rights in the basin is a premier example of the application of scientific principles to guide water law. The spirit of Morris Bien lived on.

"That Reynolds," critics would later say, "he took the Rio Grande and put it in the wells." But Reynolds's perspective was different and perhaps unique. He didn't see this as a transfer of river water but as just a reshuffling of the points at which basin water was authorized to be consumed. After all was said and done, and there would yet be a lot of debate and activity, an ocean of untouchable groundwater would remain in the middle Rio Grande aquifers. The river would also continue to flow. The compact, not environmental wisdom, required this outcome.[34]

"That Reynolds," his boosters and trusted lieutenants would later say, "kept the river whole." But what they considered "whole" was not the integrity of the ecological and natural river system that would only much later be articulated by biologists and ecologists. What Reynolds and his supporters meant by a "whole" Rio Grande was a river that had enough flow to meet the legal demands of the law of the river—the Rio Grande Compact—all the while maximizing beneficial use in New Mexico.

Reynolds Gets the River Under Control

On one front, Reynolds battled for control of the Rio Grande valley groundwater. On another, he had to get control of New Mexico's huge surface water debt under the Rio Grande Compact. In the long term, he had to allow a larger proportion of Rio Grande water to reach Elephant Butte Reservoir, but the short-term battle was a legal one. Texas had filed an interstate water suit against New Mexico. Here Reynolds's strategy was to buy time. He liked to quip, "The Supreme Court is often wrong in these water matters, but it *is* supreme." Recognizing this fact, his lawyers quickly avoided a court ruling by persuading the Supreme Court to dismiss the compact case against New Mexico and Colorado on technical grounds, putting off for the moment the threat of more federal control.[35]

Now Reynolds could turn his attention to achieving physical control of the Rio Grande itself. From the beginning of his long tenure, Reynolds supported and lobbied for annual appropriations that would allow Reclamation and the Corps to complete middle Rio Grande projects envisioned in the plan for the 1948 Flood Control Act and authorized in the Flood Control Act of 1950. The flood control plan called for myriad interrelated Rio Grande projects: the resurrection and rehabilitation of the MRGCD; the rebuilding of flood control levees along the middle reach of the river around Albuquerque; the clearing of the clogged entry into Elephant Butte; the construction of tributary dams; perhaps a main-stem Rio Grande dam to capture floods and control silt; and, last but not least, the channelization of the 147-mile reach of the river between Espanola, north of the Otowi gauge, and Elephant Butte (see plate 18).[36]

These many-year, multifaceted projects had been limping along before Reynolds took over in 1955; under his leadership, they took off. Work started simultaneously at the north and south end of the long river reach. At the south end, Reclamation first attempted to eradicate the saltcedar and other nonnative vegetation that had built up and clogged the upper end of Elephant Butte Reservoir. At the northern end of the Rio Grande valley, Reclamation dredges took a river that had meandered gracefully through the Espanola Valley and straightened and deepened it. The river, by 1954, ran in a straight, true course right through the valley, leaving on its sides the dry winding channels of its previous self-created course. What about the fish, the birds, the plants that had formerly depended on the myriad of habitat niches along the diverse meandering channels? The short answer is that nobody, or almost nobody, cared. The river now delivered water more efficiently to the irrigators downstream, and it didn't flood over the levees any more. Those were the benefits that mattered.[37]

The new state engineer shared these feelings when he took over in 1955 and, with his considerable political skills, supported the continued reengineering of the river to meet these purposes. Year after year, Reynolds worked with private and public New Mexico interests, with Congress in Washington, and with the agencies there to push for the extension of the Rio Grande Project north from Elephant Butte and south from Espanola. In the process, Reynolds formed a powerful leg of what later critics would call "the Iron Triangle" of western water: western state engineers, the Bureau of Reclamation, and long-term western senators and congressmen. Together the three legs of the Iron Triangle straightened and "improved" the channel of the middle reach of the Rio Grande.[38]

"That Reynolds," critics would later say, "took the Rio Grande and turned it into a ditch." He'd had a lot of help from people who thought like him, who believed that God had created and placed rivers so that humans could make use of them. A river that was engineered to better serve human communities was, to these people, just a better river, not a ditch. Efficiency was the goal, not preservation.

Now the problem was that the increased flows from the more efficient river couldn't reach the reservoir at Elephant Butte because the attempt to eradicate saltcedar upstream of the reservoir had failed. Clearing the delta at the head of Elephant Butte Reservoir hadn't guaranteed that water in the improved river channel would reach its storage destination. Many times when flow was low the water simply petered out before it reached the delta, leaving the sandy channel dry.

Reclamation and Reynolds came up with an even more radical solution to this problem. They constructed a huge straight ditch, its base ten or twenty feet below the bed of the Rio Grande, and then diverted nearly the entire flow of the river down this Low Flow Conveyance Channel (LFCC). Now the hydrological situation was reversed. Not only did the channel move water downstream with less loss from evaporation and streamside plants, it actually pulled water from the surrounding aquifer. During critical times of low flow, the discharge of the channel increased downstream instead of diminishing to nothing. Elephant Butte Reservoir's water level soon began to rise in response to the new inflows, and, little by little, the water debt to Texas began to shrink. However, this success came at a heavy price. The low flow channel didn't modify the river's transport system; it replaced it altogether. As time went on, many inhabitants of the Rio Grande valley began to perceive this enormous slash across the landscape as a disturbing exaggeration of the confinement visited on the upstream river. The new LFCC fit Reynolds's vision of the relationship of man and nature. It was no coincidence that work on the channel began in the 1950s, toward the beginning of his tenure, and it became fully operational in the late 1950s, at the time when his many projects started to coalesce.[39]

New Water for the River

At the same time that Reynolds was marshaling efforts to improve the Rio Grande's ability to transport water, he was also pushing efforts to increase the amount of water in it. He had always been skeptical of grandiose plans to bring additional water to the Southwest via cloud seeding, a technology on which he himself was an expert, or to pipe it from the water-rich Pacific Northwest. But the San Juan River west of the Continental Divide ran through a small part of the northwestern corner of New Mexico. This was a more convenient target and New Mexico had a legal claim to it. The problem was that getting San Juan River water under the Continental Divide would require some monumental engineering.

The state had long eyed the San Juan as a potential supplemental source of Rio Grande water. By the 1930s, water managers suspected that additional water might be needed because rights to the Rio Grande already exceeded its reliable supply. By the late 1940s and early 1950s, many New Mexicans thought that San Juan water could offset the groundwater pumping in the middle Rio Grande. By the beginning of Reynolds's term as state engineer, New Mexicans were coming to realize that they had better put their equitable share of San Juan water to some use, otherwise Arizona and California would gobble it up and New Mexico would never get it back.[40]

Despite these perceived Rio Grande needs and given the regional political realities, it took Reynolds's touch and powerful New Mexican congressmen to patch together a federally sponsored transbasin diversion. The water would be diverted from the headwaters of the San Juan River in Colorado and west of the Continental Divide, into the Chama River east of the divide, and from there into the Rio Grande. To make it work, Reynolds had to successfully negotiate with existing San Juan River users in New Mexico and Colorado; with the Navajo Nation in New Mexico and Arizona; with the other six Colorado River states who were trying to cope with the constraints of two interstate compacts on the river; and with one mammoth Supreme Court decision interpreting them.[41]

He used different strategies for different parties. For example, with regard to the Navajo Nation, Reynolds helped secure federal funding for the Navajo Irrigation Project, a development designed to bring 110,000 new Navajo acres under irrigation. To Arizona, he offered support for the controversial Central Arizona Project in exchange for support of the San Juan–Chama diversion. When Arizona congressman Mo Udall accused Reynolds of blackmailing him, the unflappable Reynolds replied, "Well, go ahead and use the term if you want to, if it helps you to understand New Mexico's position."[42] Out of deals like that, in 1962 Reynolds and his followers got Congress to authorize and fund the San Juan–Chama Project that would import 110,000 additional acre-feet of water from the San Juan basin into the Rio Grande basin.[43]

That same year, Reclamation quit working on the restructuring of the Rio Grande itself and began constructing the massive works—dams, tunnels, reservoirs—necessary to the transbasin diversion. The heart of the engineering project was three huge tunnels—five, nine, and thirteen miles in length. By 1970, 110,000 acre-feet of water per year were flowing under the Continental Divide. Pushed, cajoled, and supported by New Mexicans like Reynolds, Reclamation had transformed the capacity of the Rio Grande to carry water and now provided new water to flow into the newly engineered river.[44]

New Mexico's success in seizing a truly new supply of water to replenish the Rio Grande left flood and silt control as the major problems on the river. Under the Flood Control Act of 1948, these troubles on the Rio Grande had fallen into the purview of the Corps, which began to address them by constructing Jemez Canyon Dam in the early 1950s. The Corps then proceeded to build both the Abiquiu and Galisteo dams later in the same decade. These tributary dams were each controversial in their own right, but all were designed to help correct two problems on the Rio Grande main stem: silt aggradation and floods.[45]

None of the dams addressed the root of these problems. The increased silt had been caused by poor land-use practices in the uplands; the dams dealt with downstream consequences, not upstream causes. For flood protection for Albuquerque, the three tributary dams would help, but wouldn't be sufficient. Because these dams were all built on tributaries, the Rio Grande valley remained vulnerable to a flood pouring down the river from Colorado. Only a main-stem flood control and silt retardation dam would protect Albuquerque, New Mexico's largest and fastest-growing urban area.[46]

Since the 1940s, the Corps had been searching for a site for this purpose, sometimes near the Colorado-New Mexico border, more often farther south. In the early 1960s, the focus shifted south, much to the consternation of the five middle Rio Grande pueblos north of Albuquerque and below La Bajada. The Corps threatened to condemn land that the pueblos might not willingly relinquish. Finally, Cochiti Pueblo struck a reluctant deal with the Corps. By 1970, Cochiti Dam was a four hundred-foot-tall reality, towering over Cochiti Pueblo and the middle Rio Grande below.[47]

Reynolds had helped shepherd the necessary federal plans and appropriations through the federal bureaucracy. After the dam was built, he continued to defend the wisdom and necessity of it against mounting environmental criticism. In Reynolds's view, Cochiti Dam was the capstone of a program, begun in earnest only fifteen years earlier, to bring the entire middle Rio Grande under control.[48]

Tall dams were now poised at the top and bottom of the critical middle Rio Grande. Silt and flood dams flanked both sides of the river. These dams have permanently altered the Rio Grande. They were partly intended to reduce the flood of silt from the overexploited highlands that was choking the river channel, but they tipped the balance in the other direction, releasing water so low in sediment that the river responded by eroding its channel. Below Cochiti, the Rio Grande now flows through a deep trench, leaving the floodplain high and dry. The dams evened out the seasonal variation of flow, nearly eliminating the spring floods that once kept the channel wide and mobile while replenishing the adjoining bosque forests. Even those portions of the river that were not actually channelized now more resemble a large ditch than they do the river of a hundred years ago.

For better or for worse, all of these changes—public control of groundwater pumping, replumbing the Continental Divide, channelization of the river, big dams—are the legacy of Reynolds and the men who worked with him. To those who care about the Rio Grande for its own sake, the "worse" is the largely irreversible transformation of the Great River of the Southwest into something resembling a ditch. Reynolds and his compatriots would respond that the "better" was that the river had been retooled to most efficiently make water available for human use.

Even assuming that the goal of deriving the maximum human benefit from the river was justified, were the means inevitable? In fact, there were alternatives to reengineering the hydrological landscape. New Mexico had to look no farther than Colorado, its neighbor to the north, to see what these might be. At the same time that New Mexico was falling farther in debt to Texas, Colorado was also required to reverse its accumulated debt of water deliveries to New Mexico and Texas under the compact. Colorado was under the imminent threat of a Supreme Court lawsuit when, in 1985, nature intervened and water spilled over Elephant Butte Dam due to a series of wet years. Under the arcane rules of the compact, all accumulated water debts were erased, letting Colorado off the compact hook. Coloradans celebrated by toasting the river with champagne from the top of the spillway of Elephant Butte Dam. Thereafter, Colorado stayed the course by shutting off junior water rights whenever compact compliance was threatened. Colorado opted to control man, not nature. New Mexico did the opposite: it increased the human use of water and tried to make up for it by reengineering nature.[49]

Nevertheless, New Mexico's efforts to work itself out of its compact dilemma by controlling nature seemed to be equally successful so far as the balance of deliveries was concerned. From 1940 to 1955, New Mexico had accumulated a 500,000-acre-foot debt to Texas and southern New Mexico farmers below Elephant Butte Dam. But in 1955, the year Reynolds took over, and for reasons no one could exactly quantify, the trend reversed and for the next fifteen years New Mexico began to deliver annually more water than the compact required. Its credits ate into its deficits. This surplus delivery was achieved despite persistent below-average flows of the Rio Grande at Otowi.[50] By 1971 New Mexico was 51,000 acre-feet ahead rather than 500,000 acre-feet behind. Graphed in terms of water debt, the upward trend through 1955 and the downward trend through 1970 presented the topographical profile of a western mountain. Wisecrackers named its contours "Mount Reynolds" (see plate 19).

In the complex Rio Grande watershed, who could say which of the amazing range of programs that had come about in the first fifteen years of Reynolds's stewardship had played the biggest role in making the river perform better? The cause could very well have been that increased groundwater pumping was counteracted by purchasing old surface water rights and taking them out of use. It might have been the river's new, straightened channels that delivered river water more efficiently downstream. It might have been the increased control and release of the tributary and main-stem flood flows. Maybe it was the increased return flows to the river from Albuquerque's wastewater treatment plant, for a while at least, a new source of water for the river. Perhaps it was due to different patterns and amounts of precipitation. The one sure thing was that Reynolds created Mount Reynolds and scaled it for New Mexico. In so doing, he transformed the river.

The braided, twisting, changing channels of the Rio Grande had been remolded into one deep channel. Where prior to 1940 the Rio Grande had frequently overrun its many banks, after 1955, it remained confined in one deep, straight, man-made channel in the middle reach. The former sinuous bends had been straightened by levees that confined the river to a narrow strip of land. In a desire to further stabilize the system, the Corps had in 1953 begun to install *jetty jacks*—giant-sized versions of the pieces in the familiar children's game, made out of steel beams or heavy timbers (see plates 20 and 21).[51] By 1962, about 115,000 of these devices had been installed. Any sediment carried by overbank flows now dropped out around the levees and jacks, creating high banks that the diminished river could no longer overtop. The sediment also trapped cottonwood seed and so encouraged thick new growth of the bosque in these high banks.

Legally, and as an accounting matter, the water in the Rio Grande channel was now made up of different imaginary layers. There had always been natural base flows and natural flood flows. To those layers were now added flood flows captured temporarily and released from Corps tributary and main-stem dams; stored flows released for the MRGCD; imported flows from the San Juan–Chama transbasin diversion; and return flows from the City of Albuquerque. Every cubic foot of water released into the Rio Grande was owned by somebody, and it was intended to be delivered to somebody else. A freight train may appear from a distance to be a long, continuous, moving snake, but in reality it consists of a sequence of distinct objects that are being transported from the possession of a previous owner to a new owner. The Rio Grande was no longer the ecological bloodstream of a desert valley, it was only a legal freight train that existed for the purpose of delivering parcels of water from one owner to another.[52]

Every layer or parcel of water carried a different legal claim and different legal baggage. Between 1945 and 1970, the Rio Grande became a river governed by human law rather than natural law. In the process of remaking the river to meet the compact, Reynolds damaged its capacity to perform as a river. Any attempt to reset that balance would have to await a new political and legal climate.

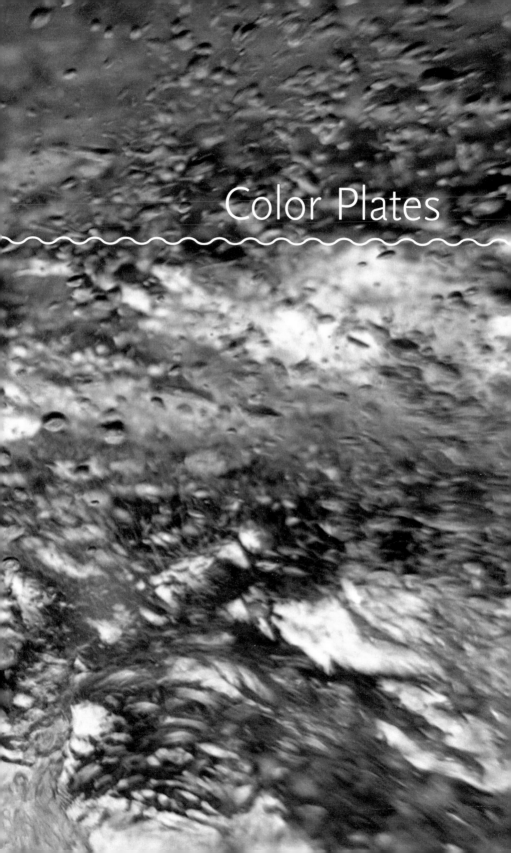

Color Plates

Plate 1
Today, the imposing sweep of Cochiti Dam divides the upper and lower Rio Grande.
Cochiti Pueblo is at bottom of picture. Photo courtesy of Thomas Blog.

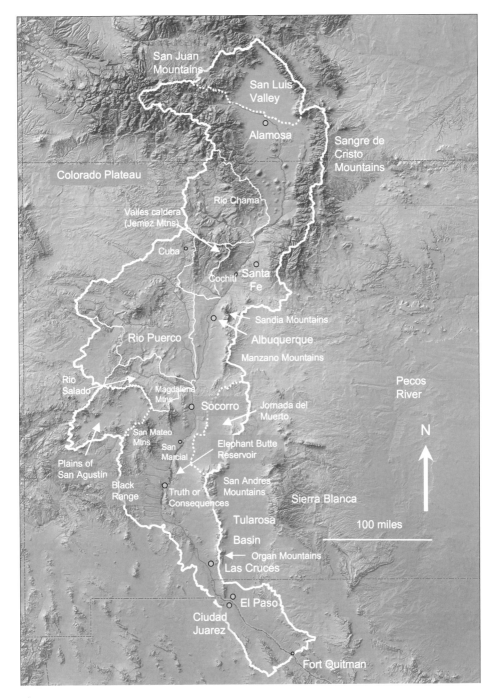

Plate 2
Drainage basin of the upper Rio Grande. Tributary basins that are connected to the Rio Grande are indicated by thin solid lines. Hydrologically closed drainage basins associated with the Rio Grande are indicated by dotted white lines.

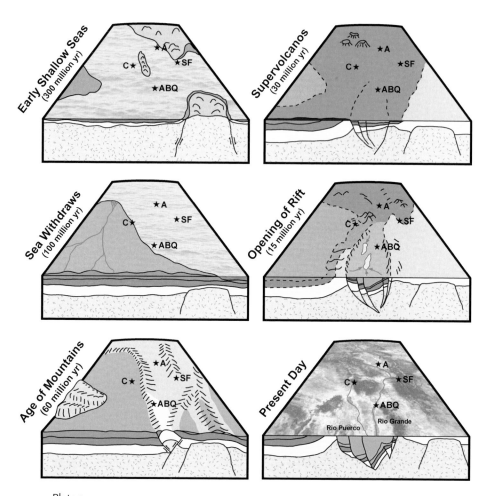

Plate 3

Geological history of the northern portion of the upper Rio Grande. The view is from over the Rio Grande between present-day Albuquerque (ABQ) and Socorro looking north. "SF" is the future location of Santa Fe, "A" is Alamosa, and "C" is Cuba. The pink color is the Precambrian basement, blue is mostly marine sediments (especially limestone), gold and green are mixed marine and continental sediments, red and orange are lava, and yellow is rift-fill sediment. Images by Fred Phillips, Mike Buffington, and Susan Delap-Heath.

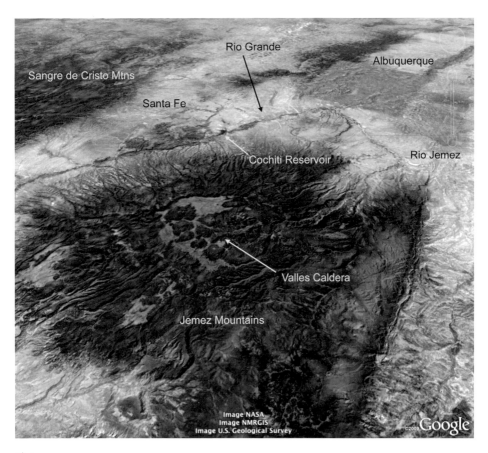

Plate 4
A view of the Valles Caldera in the Jemez Mountains from the west. The crater (caldera) left by the supervolcano eruptions of 1.6 to 1.2 million years ago is the semicircular light area in the center of the photo; the dark areas surrounding it are mostly lava from those eruptions. Photo courtesy of GoogleEarth.

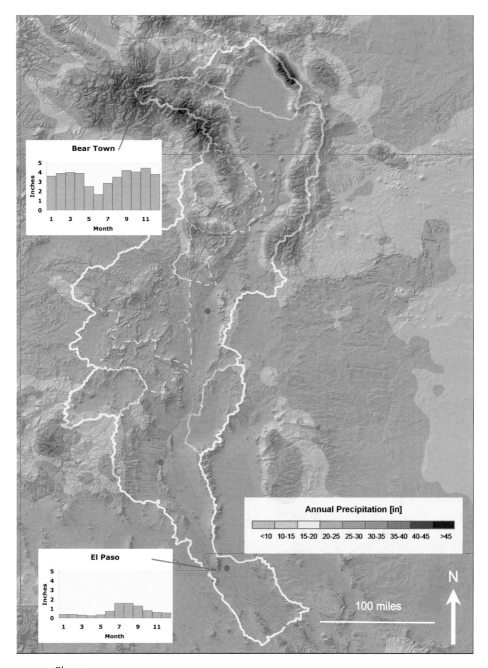

Bear Town

El Paso

Annual Precipitation [in]

<10 10-15 15-20 20-25 25-30 30-35 35-40 40-45 >45

100 miles

N

Plate 5
Average annual precipitation over the Rio Grande basin. The two graphs illustrate the
difference in the seasonal pattern of precipitation between the northern and southern
portions of the basin.

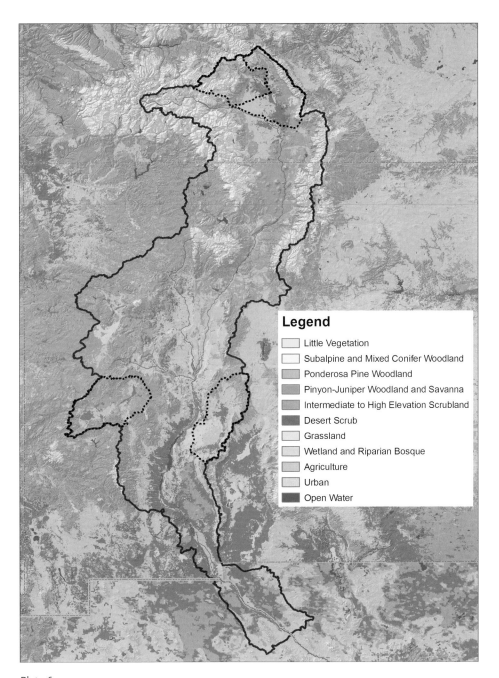

Legend
- ☐ Little Vegetation
- ☐ Subalpine and Mixed Conifer Woodland
- ☐ Ponderosa Pine Woodland
- ☐ Pinyon-Juniper Woodland and Savanna
- ☐ Intermediate to High Elevation Scrubland
- ☐ Desert Scrub
- ☐ Grassland
- ☐ Wetland and Riparian Bosque
- ☐ Agriculture
- ☐ Urban
- ☐ Open Water

Plate 6
Distribution of plants in the Rio Grande basin. Plant communities characteristic of humid regions (e.g., pine forest) are found mainly at high elevations while those associated with dry areas (e.g., creosote scrub) grow at low elevations. Image courtesy of Matej Durcik and Fred M. Phillips.

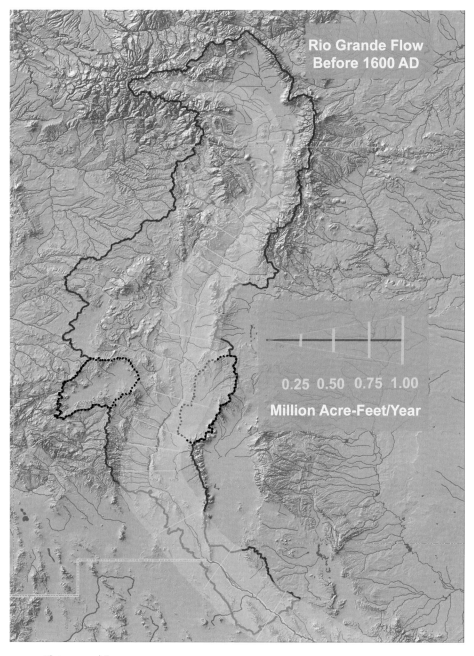

Plates 7 and 8
A comparison of the average annual flow of the Rio Grande before extensive human development (about AD 1600) and today. The width of the light blue lines is proportional to the amount of flow in the river. The contemporary river discharges are based on USGS, Reclamation, and International Boundary and Water Commission gauging data (Mills 2004, 49, 77). Predevelopment discharges were reconstructed by adding back in contemporary net diversions (Mills 2004; S. Ellis et al. 1993). Images by Matej Durcik and Fred M. Phillips.

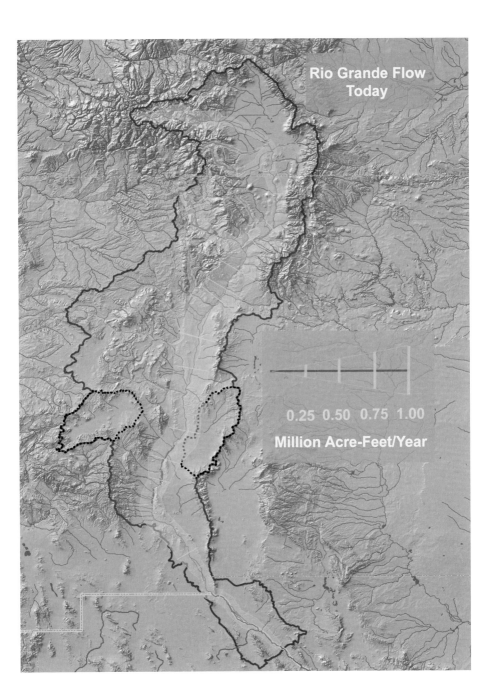

Rio Grande Flow
Today

0.25 0.50 0.75 1.00
Million Acre-Feet/Year

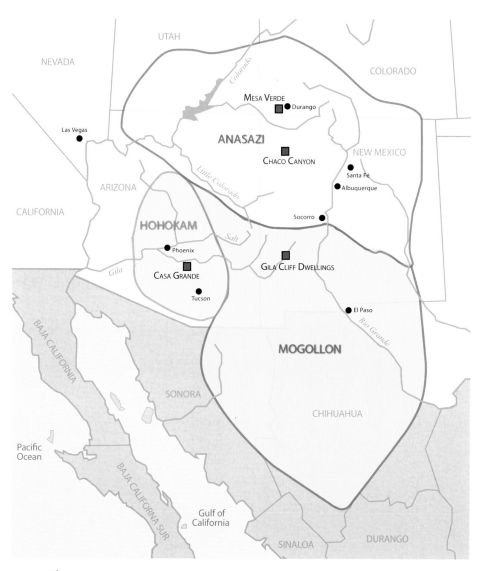

Plate 9
Map of the extent of Anazasi, Hohokam, and Mogollon settlements. Wikipedia file
Anasazi.svg. Modifications by Arkyan.

Plate 10
The precise masonry construction accomplished by Chacoan architects and builders is reflected in this series of doorways that have well endured the passage of time at Pueblo Bonito. Photo by John Villinski/abstractsouthwest.com.

Plate 11

A satellite view of the village of Cundiyo, looking south toward the irrigated fields that lie upstream of the village. The Acequia del Molino runs at the eastern edge of the irrigated fields and flows toward the site of the old mill, where the valley narrows down close to the village. Figure based on a GoogleEarth image.

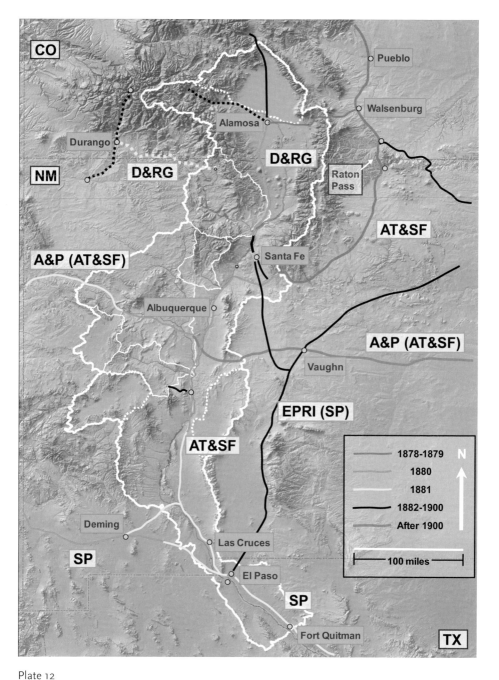

Plate 12
Railroad construction in the Rio Grande basin. The timing of construction is shown by colors. Narrow gauge railroads are in dashed lines. AT&SF = Atchison, Topeka, and Santa Fe; SP = Southern Pacific; D&RG = Denver and Rio Grande; A&P = Atlantic and Pacific; EPRI = Rock Island and El Paso. Only railroads in the vicinity of the Rio Grande basin are shown, and most local lines are omitted. Image courtesy of Matej Durcik and Fred M. Phillips.

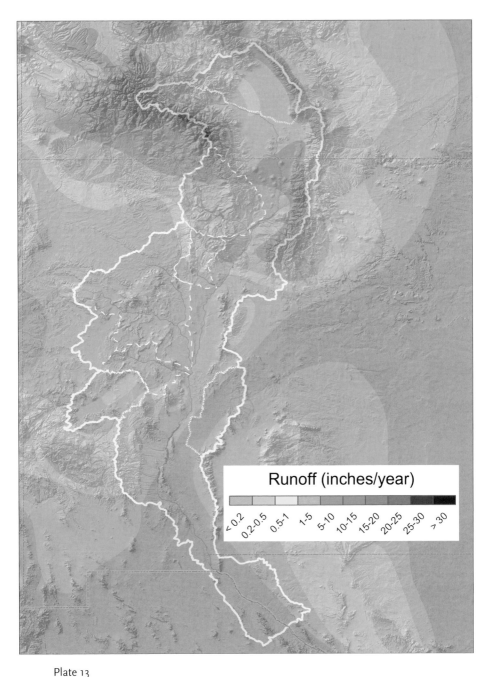

Runoff (inches/year)

< 0.2 0.2-0.5 0.5-1 1-5 5-10 10-15 15-20 20-25 25-30 > 30

Plate 13
Runoff produced by precipitation over the Rio Grande basin. The units (in/yr) show the
amount of precipitation that must run off to supply the flow in the streams draining each
portion of the basin. Nearly all of the areas that produce large amounts of runoff are
found at the highest elevations, and the monster canal diversions in the San Luis Valley
cut off the areas producing the most flow from the rest of the drainage basin. Image
courtesy of Matej Durcik and Fred M. Phillips.

Plate 14
Delegates' badge and postcard from the Sixteenth National Irrigation Congress, held in Albuquerque in 1908. The inscription on the badge reads "Science bids the desert drink." Photo by Fred M. Phillips.

Plate 15
The late Steve Reynolds stands astride the Santa Fe River in typical attitude. Photo © by Danny Turner. All rights reserved, DannyTurner.com

Plate 16
A new channel cuts straight through the meandering course of the Rio Grande near
Espanola, 1957. Photo Courtesy of Bureau of Reclamation Albuquerque Office.

Plate 17
Waterlogged lands along Rio Grande Boulevard in Albuquerque, July 1, 1930, near the site of the Nicolas Sanchez farm.

Plate 18
The same land in spring 1933, after drainage and restoration by the MRGCD. Photos courtesy of MRGCD.

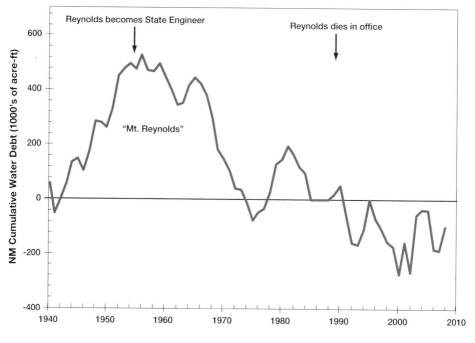

New Mexico Water Debt under the Rio Grande Compact

Plate 19
Data from the New Mexico Interstate Stream Commission (2010).

Plates 20 and 21
Installation of jetty jacks in the Albuquerque reach of the Rio Grande shows a dramatic effect on bosque. Top picture was taken just after the jacks were installed in April 1960; bottom picture was taken in May 1972, and shows significant infill of bosque just outside the line of jacks. Photos Courtesy of Bureau of Reclamation Albuquerque Office.

Plate 22
The famous "Blue Marble" photo taken on December 7, 1972, by astronaut Jack Schmitt as Apollo 17 traveled to the moon. Astronaut photograph AS17-148-22727 courtesy NASA Johnson Space Center.

Plate 23
The tiny silvery minnow, focus of so much legal and environmental attention in the Rio Grande basin, is dwarfed by the palm of a hand. Photo by Aimee M. Roberson, USFWS.

Plate 24
The New Mexico Interstate Stream Commission's Los Lunas Silvery Minnow Refugium, shown being filled in April 2008, features a 458-foot-long meandering stream and mimics the natural flow regime, including spring snowmelt runoff and associated flooding. Photo by Douglas Tave and the New Mexico Interstate Stream Commission.

Plate 25

The Los Lunas refugium shown from the opposite perspective, in 2009, after filling and plant growth. Photo by Douglas Tave and the New Mexico Interstate Stream Commission.

Plates 26 and 27
The bosque in Santa Ana Pueblo before and after restoration efforts successfully cleared the underbrush. Photos courtesy of the Pueblo of Santa Ana.

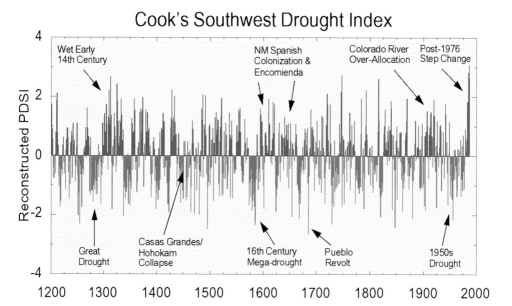

Plate 28

This drought chronology, based on the tree ring record and a reconstructed Palmer Drought Severity Index (Cook 2000), shows an interesting correspondence of periods of significant social upheaval and drought in the southwestern United States. Events labeled by Julio Betancourt.

Historical

1953-1956

2000-2003

Future

2006-2030

2035-2060

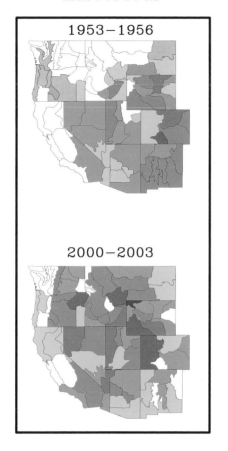
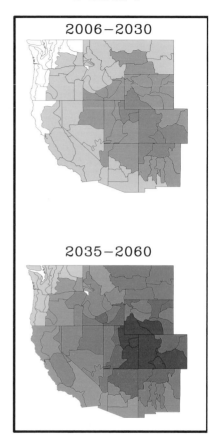

PDSI

-6 -5 -4 -3 -2 -1 1 2 3 4 5 6

Plate 29
Palmer Drought Severity Index. The figures at left show past drought conditions in the West (values less than -3 denote severe drought conditions). Those on the right show the Intergovernmental Panel on Climate Change simulations for periods through 2060. Note that the San Juan Mountains, the source of the Rio Grande, are at the bull's-eye of the future drought region. According to these predictions, by around 2050 the "average moisture balance conditions will mimic conditions experienced only rarely at the height of the most severe historical droughts.

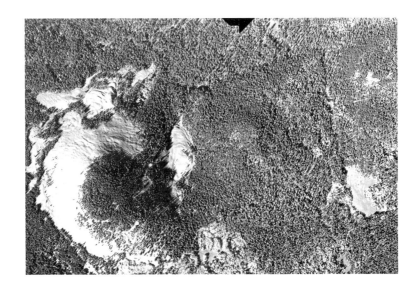

Plates 30, 31, and 32
Aerial photography of Redondito Peak in the Valles Caldera National Preserve showing logging impact on land cover prior to logging in 1963; after logging in 1975 (showing logging roads); and near-closure of dense second-growth canopy in 2005. Photos courtesy of Valles Caldera Trust.

Plate 33
Water builds up behind a diversion dam constructed on the Rio Grande as part of the
San Juan–Chama Drinking Water Project in Albuquerque, just before its opening on
December 4, 2008. AP/Wide World Photos/Susan Montoya Bryan.

Chapter IX

Shifting Values, New Forces on the Rio Grande

As He Lay Dying

In April 1990, State Engineer Steve Reynolds returned to Santa Fe from mopping up yet another water crisis in San Juan County, New Mexico. As a severe and long-time diabetic, Reynolds knew when he was slipping into insulin shock. As soon as he got home, he admitted himself to the St. Vincent Hospital. He never emerged. He had said over and over again that he would die with his boots on and he did, one week later.

To his way of thinking, he had left the Rio Grande water world in pretty good order. He recently had personally outmaneuvered El Paso, Texas, in its efforts to develop New Mexico Rio Grande basin groundwater for its own use.[1] Over the thirty-five years of his rule as state engineer, he had done everything in his power to control the Rio Grande so that the river would meet its international and interstate obligations and still leave as much as possible for New Mexico's economic development.

Despite his significant accomplishments, Reynolds had overlooked late in his career new forces that were slowly gathering strength. In the late 1960s and early '70s, a new perspective started to grow around him. On December 7, 1972, New Mexico's own geologist-astronaut Harrison H. ("Jack") Schmitt, an *Apollo 17* crew member born and raised in Silver City, took the famous "Blue Marble" picture of Earth (see plate 22) that showed it to be a tiny blue-and-white-flecked jewel suspended in space.[2] The picture hit home: this was it. There was no place else to go, and we had better take care of our planet. Paul Ehrlich's *Population Bomb* and Rachel Carson's *Silent Spring* had already brought national attention to the fact that there might be limits on the size of sustainable communities and the quality of the water available to support them. National environmental groups such as the Sierra Club, its local New Mexico chapter, and local offshoots like Santa Fe's Central Clearing

House began to focus attention on the harmful ecological and watershed impacts of Reynolds's Rio Grande water policies. When the Central Clearing House awarded Reynolds its annual Earth Enemy of the Year Award in 1980, Reynolds, increasingly out of touch but as wry as ever, claimed that, of the many awards he had received, this was the one of which he was most proud.[3]

Besides the growing New Mexico environmental movement, other new interest groups sprouted and thrived. Hunters and fishermen in New Mexico had been around for as long as New Mexico had been a state.[4] But now they shared a common interest with increasing numbers of others whose recreational enthusiasms were also water based and water dependent, but focused on activities such as boating, birding, and backpacking. Since the 1960s, summer campers and boaters enjoyed Elephant Butte Reservoir on the Rio Grande main stem and Santa Cruz Lake on one of its northern New Mexico tributaries, as well as the Rio Chama reservoirs. In the mid-1970s, with the help of the increasingly powerful New Mexico senator Pete Domenici, Cochiti Reservoir underwent a transformation, with a change in the dam's status from a flood control dam (which allowed storage of water for no more than seventy-two hours) to a dam with a permanent reservoir.[5] It became a recreational hot spot, just forty-five minutes from both Albuquerque and Santa Fe. Late in his tenure as state engineer, Reynolds even bought himself a small sailboat and hauled it behind his ancient and beloved Mercedes Benz to Cochiti Lake to sail alone on hot summer Sunday afternoons.

Demographic shifts that transformed the Rio Grande valley from an agricultural to an urban economy shifted some of the political balance of power when it came to water. The booming Golden Triangle formed by Los Alamos, Santa Fe, and Albuquerque became a hotbed of the new water values in the middle Rio Grande.[6] The rights to use water in New Mexico remained private property rights, more than 80 percent of which were devoted to agricultural uses.[7] But the water itself remained public, sharply increasing the tension between the agrarian holders of private water rights and the growing urban populations who valued water as a public resource for nonagrarian uses.[8]

Intellectual shifts contributed to further undermining the basic stance of "water buffaloes," the slight pejorative applied to long-time state administrators like Reynolds. People of his ilk always had believed that in a perfect world water would flow uphill to money, meaning that water rights would rise through water markets to their highest, best, and most efficient economic uses, usually from agriculture to municipal and industrial uses. This vision, however, did not embrace many of the ways in which human beings need or value water. Economists, biologists, and hydrologists had begun to assign economic value to nonmarket uses of water, to the ecosystem services that rivers like the Rio Grande provided.

Reynolds shuddered when the new economists spoke of the money value of recreation or of the existence value of the Rio Grande. How much stranger he must have found talk of wetlands as the indispensable kidneys of a watershed or of floodplains as themselves providing flood protection. This kind of ecosystem lingo swirled around him late in his life and complicated the question of the value of the Rio Grande itself.

In Reynolds's final decades in office, the federal government rode the wave of these ideas back into supremacy over the states in water matters. Held at bay for years by westerners like Reynolds, the federal government reasserted its power over natural resource issues with two major federal acts passed just a year apart: the Clean Water Act of 1972 (CWA; amended in 1977) and the Endangered Species Act of 1973 (ESA).[9] The CWA addressed problems of the environmental degradation of surface water, aiming to reduce toxins introduced to rivers like the Rio Grande by nonagricultural uses. It authorized states or tribes to adopt and enforce water-quality standards for rivers like the Rio Grande. New Mexicans learned what that meant in the 1990s when the Pueblo of Isleta, just downstream from Albuquerque's wastewater discharge, adopted water quality standards that neither the city nor any other entity nor nature itself could meet.

In parallel with the CWA, the ESA strove to protect and recover threatened and endangered species and protect their habitat. Since three-quarters of the vertebrates in New Mexico and Arizona depend on riparian habitats,[10] the Rio Grande was bound eventually to become ground zero for the newest battle: the river as habitat for nonhumans.

The seeds for all of these changing values and forces were planted in the last decade of Reynolds's long regime. He didn't like them and he didn't deal with them. But after he died in April 1990, these seeds started to take root, emerge, and even bloom in a tangled garden of new, different, and chaotic growth.

A Professor Considers the Rio Grande Bosque

Biology professor Cliff Crawford regretted his decision. In September 1965, the Rio Grande bosque in the middle of Albuquerque had seemed like a logical place to lead a field trip for his University of New Mexico entomology class. But now that the students were piling out of the vans, Crawford wished he had scouted the location first. Students were tripping over trashed appliances and abandoned automobiles. The stench from the rotting piles of illegally dumped garbage competed with clouds of mosquitoes for dominance of the hot, humid air. Where movement was not hindered by trash, thickets of saltcedar blocked access to the riverbank. After an unpleasant half hour or so, Crawford herded the class back into the vans. He would not be tempted to come back to the bosque for many years.

In 1986 Crawford returned from a sabbatical partly spent in the hyper-arid deserts of Namibia in southern Africa. There he had examined with fascination the ecosystems of rivers that made their way through the barren sand dunes of the coastal desert. On arriving home, it struck him that he had his own desert river to study, one about which he knew almost nothing. He revisited the Albuquerque bosque and was pleasantly surprised to see that much of the trash had been hauled away and the access roads blocked so that more could not be easily added. Hikers and joggers were beginning to use the river corridor on a regular basis. In 1988 he combined forces with Manuel Molles, another biology professor from the University of New Mexico, to offer a new course on bosque ecology.

The course was more accurately an ongoing research project, since not much was actually known about the functioning of this riparian system. Why had invasive saltcedar and Russian olive taken over much of the bosque? How did their water use compare with that of the native cotton-wood and willow? How had confinement of the river between levees affected aquatic species such as native fishes? Many such questions were unanswered. Gradually, over a span of years, Crawford, Molles, and their students tried to figure out why the river barely resembled the one the Spanish first reported on in the mid-sixteenth century.

Biological diversity had declined dramatically. Much of the native vegetation was choked out by invaders. Half the original twenty-four species of fish, including sturgeon, gar, and eel that had once migrated up from the Gulf of Mexico to breed were gone.[11] Beavers and mink that had inhabited the wetlands had disappeared, as had wild turkeys and whooping cranes. The remaining birds and mammals had declined to insignificant numbers. The riparian zone of the Rio Grande was a dying habitat.

The timing of Reynolds's death coincided eerily with a groundswell of support for attending to ecological concerns when managing the river, almost as if his passing had released a quiet underground revolution. In the late 1980s, managers of the Bosque del Apache National Wildlife Refuge had decided that its mandate went beyond providing shelter and food for migrating ducks and geese. They began to rip out invasive species and encourage native vegetation in its place. By 1990 visitors were impressed with the results and began to wonder whether something similar might be done on a larger scale. The U.S. Army Corps of Engineers became interested in cottonwood regeneration.[12] The Bureau of Reclamation began to adjust its releases on the Chama River to promote brown trout spawning on the upper reach of the river between El Vado and Abiquiu reservoirs. And most important, word of these developments reached the ear of Senator Domenici, who in 1991 encouraged the formation of a Rio Grande Bosque Conservation Committee to recommend how the bosque might be returned to something resembling a functioning status.

The committee recommended that qualified experts put together a management plan for the bosque. To be successful, any such plan had to get buy-in from the major management agencies, so the panel included representatives from Reclamation, the Corps, and the U.S. Fish and Wildlife Service, among others. The committee was headed by an independent expert, none other than Crawford. In 1993, the panel issued a landmark plan for the management of the middle Rio Grande bosque.[13]

What did bosque restoration really entail? Was this a plan to turn back the bosque clock to the sixteenth century, before humans had undertaken major interference with the system? Crawford's answer:

> Given the changes in the Middle Rio Grande Valley that were caused in this century by regulation and other forces (including colonization by introduced woody plant species), a return to the pre-settlement condition is clearly impractical. Under the circumstances, we therefore advocate the reestablishment of basic riverine-riparian functioning rather than the "saving" of a bosque that is itself an artifact of civilization. We propose that sustaining ecosystem integrity, in the form of carefully planned partial restoration, is the only reasonable alternative to irreversible ecosystem change.[14]

The essential elements of the management plan consisted of attempting to reintroduce at least a muted version of the ancient spring floods that transported and nourished the seeds of many bosque species; to manage the riparian forests to encourage habitat diversity; and to create wetlands both inside and outside the levee system. In addition, they recommended a sustained program of monitoring, research, and education in and about the bosque.

From the plan arose the Middle Rio Grande Bosque Initiative, an interagency ecosystem management effort that still spends about $500,000 of federal money per year for projects to restore habitat, modify the river functions, monitor and research the system, and educate and involve the public. Around this initiative a bewildering variety of task forces, study groups, management teams, and citizen activist groups has sprung up. Just keeping up with the slate of meetings, workshops, project activities, and policies is nearly a full-time job.

By the mid-1990s the Rio Grande bosque had undergone a revolution since that day in 1965 when Crawford hurried his class out of the festering trash pit that the Albuquerque bosque had become. Hiking, jogging, and bicycle trails now lined the river. Picnic areas and nature study centers dotted the valley near populated areas. Photographers and bird watchers descended on the bosque during the weekend. Management agencies like the Corps and Reclamation began to employ ecologists and hydrologists whose principal job was to return the river to a more natural condition. This focus has continued to the present, even while the agencies acknowledge the truth of Crawford's assertion that the river can never be returned to anything much resembling its pre-sixteenth-century state.

The really important change accomplished during the past two decades has been a change in attitude. The very large amounts of time, effort, materials, and dollars invested in upgrading the bosque habitat since Reynolds's death demonstrate that the entire river system—including its hydrology, chemistry, soil, vegetation, and wildlife—is now starting to be viewed as an integral and necessary component of sustainable life in the Rio Grande valley. A fundamental change in society's values since 1990 has driven this shift. The river is finally beginning to be seen again as something with intrinsic value. After 150 years in which it was viewed as a means to an economic end, a threat to property, or a commodity, many now see it as possessing its own worth and beauty. A whole series of separate activities at the end of the twentieth century and the beginning of the twenty-first started this quiet revolution.

Letty Belin and the Silvery Minnow

The path to an even partially restored Rio Grande was far from straight. The physical river and its riparian vegetation were inextricably linked to the habitat it provided for diverse populations of birds, fish, insects, mammals, and amphibians. The new regulatory agenda of the early 1970s aimed at protecting and preserving them, but it took some halting starts for the initiatives to gain a foothold. The many contradictory goals of river preservationists, engineers seeking hydraulic efficiency, farmers, politicians, and developers eager to satisfy growing municipal and industrial water demands would not be easily reconciled.

In the mid-1980s, Letty Belin moved to New Mexico from San Francisco, where she had worked as an attorney on California water law. She soon joined the office of the New Mexico attorney general (now U.S. senator) Tom Udall. As the state's chief legal adviser, Udall was interested in restoring some of the control over resources that had fallen to the state engineer's office and appointed Belin as the head of a new unit to do so. In the process, Belin helped find a way to use federal law to push Rio Grande restoration forward.

As assistant attorney general, Letty Belin took an immediate interest in New Mexico water rights and particularly the Rio Grande. At that time, the burning New Mexico issue was whether the state and its agencies would recognize the actual presence of flowing water in a river, or *instream flows*, as a possible beneficial use. Essentially, such a right would guarantee water for the rivers themselves. When she looked into it, Belin discovered that every state in the intermountain west recognized instream flow rights except New Mexico.[15] But New Mexico's emphasis on beneficial use was bound to make the state at least reluctant to recognize a legal right to leave water in a river as important as the Rio Grande.

Belin wrote an elegant opinion saying that the State of New Mexico could recognize an instream flow right if it wanted to.[16] (Her opinion followed on the heels of the state engineer's opinion which, remarkably, also recognized the possibility of instream flow rights but hemmed it in with difficult technical requirements.) Although Belin's opinion had little legal force, it caused an immediate flap: Opponents derided it. Farmers scoffed at it. Clearly, the State of New Mexico was not Belin's willing client, even if it was her employer. So she quit the attorney general's office and found the client she needed: a tiny Rio Grande fish.

In early 1990 Fish and Wildlife proposed to list the Rio Grande silvery minnow as an endangered species under the still-young ESA. Once endemic to the whole Rio Grande, the two-inch minnow (see plate 23) now only lived in a small part of the river and was almost extinct. It took four years for Fish and Wildlife to list the minnow as endangered, but once it did, the minnow's need for water came first under federal law. State-based water rights would have to give way.[17] In essence, centuries of New Mexico's legal structure were leapfrogged and made obsolete by the ESA.

The minnow needed precisely what the proponents of instream flow rights on the Rio Grande had argued for in vain. It required high runoff flows each spring to reproduce and needed a wide floodway with eddies and pools to mature. The minnow also required a constant flow of unimpeded water in the river so it could move freely back upstream after spawning. In short, the silvery minnow required just the kind of river that the Rio Grande had been before it was reined in by Reclamation and the Corps, with the blessing of Reynolds. From this perspective, the engineering improvements that were triumphs in the eyes of some, allowing New Mexico to scale Mt. Reynolds, were to others ecological disasters that threatened the silvery minnow.

These marvels or horrors, depending on the beholder, included: the Cochiti, Jemez, and Galisteo dams, which harnessed the spring floods that the minnow required; a narrow river confined within levees, which prevented the wide floodplain that the minnow needed; and unpassable MRGCD diversion dams, which kept the minnows from having full run of the river.

Negotiations between the United States, New Mexico, and a consortium of environmental nongovernmental organizations sputtered along through the 1990s without much progress. Then a near-catastophic fish kill spurred Belin to action. In the extremely dry summer of 1996, thousands of minnows died as the river dried and irrigators including the MRGCD took what they could from the paltry remaining flows. On behalf of the silvery minnow, the principal plaintiff, Belin, finally filed suit in November 1999.[18]

Her complaint alleged numerous violations of the ESA, but the one that emerged as the most consequential claimed that the United States had both the power and the duty to release Rio Grande water stored upstream for the benefit of the minnow, even if the releases took stored water bought and paid for by Albuquerque and the MRGCD. Both the city and the MRGCD were counting on as much San Juan–Chama water as they could get. Belin's lawsuit asserted that the United States had the power and the duty to use that water for the minnow instead. The threat posed to the water supply by this claim, as well as to the primacy of a hundred years of state water law development, inspired vehement opposition on the part of the city, the district, and the State of New Mexico. Belin and her partner were outnumbered ten to one by her opponents' lawyers.

Belin was convinced that her legal position was correct as a matter of law. She found legal support in a series of opinions from California. Surprisingly, she also found significant support from Judge James Parker, a relatively conservative Republican appointee to the U.S. District Court for the District of New Mexico, who had a strong business background and had never before shown much interest in Rio Grande water matters. Again and again between 1999 and 2002, Parker ordered the United States to release San Juan–Chama water contracted for the City of Albuquerque and the MRGCD to avoid desiccating the middle Rio Grande and precipitating a silvery minnow kill as severe as in 1996. Parker's final injunction in September 2002 triggered an immediate appeal to the U.S. Court of Appeals for the Tenth Circuit by all defendants.[19]

By then the legal issues were already drowning in political posturing. Albuquerque mayor Martin Chavez's legitimate concern was to secure the San Juan–Chama allocation as the principal source of his city's drinking water. But to hear Chavez tell it, Parker and the environmentalists were elevating the needs of fish over the drinking needs of the city's children.[20]

Environmentalist claims were equally loaded with hyperbole. To raise money, Wild Earth Guardians, for example, trotted out Wyatt the Wolf, a figment of their anthropomorphic imaginations, to speak for the Rio Grande. The iconographic Wyatt asked contributors via e-mail to urgently rescue a Rio Grande that in fact hadn't run free for hundreds of years, a river they claimed was once so full of countless fish that they "made the river flash silver against the glow of the setting sun."[21]

Amid the heightened rhetoric, the court of appeals affirmed on June 19, 2003, Parker's injunction requiring the federal government to use San Juan–Chama water first for the minnow and only second for the city and irrigation district. By then, however, the wider world of New Mexico water politics had engulfed the strictly legal issue of water for the river and its fish. Governor Bill Richardson, Senator Domenici—along with New Mexico's entire congressional delegation—and the City of Albuquerque all agreed that this time the courts had gone too far.[22]

The issue disappeared for a little while in the smoke and mirrors of Washington politics, then reappeared as riders to the annual Energy and Water Appropriations Bill. In 2004 and continuing thereafter, the temporary riders became permanent: San Juan–Chama water could not be used for endangered species in the Rio Grande.[23] The issue of whether Reclamation as the largest wholesaler of water in the Rio Grande basin had the discretion and the obligation under ESA to take care of the silvery minnow was never finally resolved in Reclamation's ensuing plans for the river. Another five years of bitter, extensive, and expensive litigation elapsed. In April 2010 the Court of Appeals for the Tenth Circuit weighed in with a second decision, this one more than a hundred pages long.[24] Packaged in the extremely arcane language of federal civil procedure, the court essentially voided the previous ten years of good faith efforts to decide what the modern law of the river would be. It was as if the lawsuits never happened. The Rio Grande was thrown back to the profound legal uncertainty that plagued it in the 1990s.

Despite the outcome, Belin and her client, the silvery minnow, did win some concessions for the river. Everyone agreed voluntarily to try to keep the river wet. The MRGCD began to emphasize irrigation efficiency and reduced its diversions by more than 25 percent. Fish and Wildlife found a new home for the minnow, hundreds of miles downstream, near Big Bend National Monument in Texas, an area free of the development pressures of the Albuquerque reach of the river.

There the minnow could survive even though it still couldn't survive in the New Mexico river that had birthed it. The irony wasn't lost on Belin. She had won the battle to save the minnow but lost the battle to use the ESA to save the river. Still, the river was a little better off after all the lawsuits and the politics. To someone of Belin's optimistic nature that was something. "It's still a question," she now says, "of whether the glass is half empty or half full." Most of the time she prefers to see it as half full. But she also understands that half a western river isn't much of a river at all.[25]

A Minnow Reprieve: The Refugia

The ESA and the Rio Grande silvery minnow, with Belin helping it along, forced river managers in the early 2000s to consider restoring riparian areas to the earlier, more pristine condition in which the endangered minnow had thrived. On the Rio Grande itself, some of these efforts were doomed to fail because the river was already so thoroughly altered, managed, and controlled that it couldn't support the minnow. One solution was to create *refugia*, areas set aside or artificially constructed to provide suitable habitat for endangered species, in this case to provide the minnows an interim stand-in for a natural Rio Grande. In refugia, the silvery minnows could be bred to augment their dwindling populations in the middle Rio Grande and also to provide new populations to reintroduce to suitable habitats outside this stretch of the river, as far away as Big Bend National Park in Texas.

To breed and propagate the beseiged minnow for the Rio Grande, river managers have constructed two artificial refugia: BioPark Refugium in Albuquerque, collaboratively established by New Mexico's Interstate Stream Commission, the City of Albuquerque, and Fish and Wildlife; and Los Lunas Silvery Minnow Refugium constructed by the Interstate Stream Commission twenty miles south of Albuquerque (see plates 24 and 25). (Another refugium is in the works by Reclamation in the southern bosque of Albuquerque.)

The refugia provide minnow breeding habitats that more closely resemble prehistoric conditions on the Rio Grande than anything the river currently can offer. The $1.2 million Los Lunas refugium, for example, offers a Disney-esque simulacrum of a 458-foot-long meandering stream, shallow sandy beaches, quiet pools, and advanced hydraulics to keep the system running smoothly, including the capacity to increase flow velocities and adjust water levels to simulate overbank flooding.[26]

Bucolic as it sounds, Los Lunas Silvery Minnow Refugium is entirely artificial. It is a collection of pipes and pumps and artificial flow channels. The water necessary to drive it was carefully and dutifully transferred to the structure from nearby wells on the grounds of an old public hospital. The refugium is fenced off to seal the expensive, computer-driven system off from predators, human and other. It represents a triumph of technology as a means to successfully breed and subsequently reintroduce an endangered species into a natural area—in this case, a reach of the Rio Grande in Big Bend National Park in Texas.[27]

On May 19, 2009, officials gathered under a large white tent near the refugium to dedicate it; be entertained by dancers and drummers from San Felipe Pueblo Elementary School; and listen to speeches, which all spoke of the amazing capacity of the refugium to mimic the disappearing natural conditions that once allowed the minnow to thrive. Dr. Benjamin Tuggle, regional director of Fish and Wildlife, stressed that the goal of the refugium was "not to save the fish, but to save New Mexico culture from itself."[28]

Restoring the River: Saving the Low Flow Conveyance Channel

One effect of the controversial, bitter, and ultimately ambiguous Rio Grande silvery minnow litigation was to focus the efforts of the previous decade to restore the Rio Grande on the specific problem of protecting its minnow. Following the 1994 decision to list the minnow, various state and federal agencies realized that they had better formulate management plans that included the minnow, or the courts would do it for them. In 1996 the agencies put out a white paper[29] that was soon matched by a "green paper" issued by the environmentalists.[30] The two groups began common discussions in a series of what they called the Green-White Group Meetings. After much back-and-forth, in 2003 these meetings morphed into the Middle Rio Grande Endangered Species Collaborative Program (MRGESCP), which was charged with the daunting task of figuring out how to spend a limited and insecure amount of federal funding, pieced together year to year by Senator Domenici, to improve the Rio Grande as habitat for endangered species. By 2009 the MRGESCP funded more than sixty programs to restore the Rio Grande and its riparian habitat. Perhaps the most curious and ironic of the programs involved the Low Flow Conveyance Channel in the reach of the river below the San Acacia Diversion in Socorro County.

In April 2009 Reclamation's Steve Hansen, deputy director of the Albuquerque office, led a reconnaissance trip to assess the current state of the LFCC. To get to the MRGCD's San Acacia diversion, the southernmost of the district's four massive Rio Grande diversions, Hansen took the back route and followed a bewildering maze of dirt roads starting north of Socorro. He knew the roads like the back of his hand.

Hansen and his companions bounced in his truck east across the barren, high desert between Interstate 25 and the MRGCD's Socorro main canal.

Once past that canal they entered verdant fields of alfalfa that stretched as far north and south as they could see. At the eastern edge of the fields stood a small bosque of cottonwoods. Passing through the copse and continuing down another dirt road, they came suddenly upon the four principal man-made features of this lowest portion of the middle Rio Grande. From west to east these were: the Socorro main canal; the LFCC inside it; a large levee; and then the Rio Grande itself, straddled by a massive brown diversion dam that ran bank to bank, capable of controlling the full flow of the river at that point, the initials MRGCD carved deep in its cement face.

Of these, the LFCC has played the most complex role in the Rube Goldberg world of Rio Grande water. Capable of carrying up to 2,000 cubic feet per second of water, often more than is in the Rio Grande at that point, and sending it swiftly and efficiently into Elephant Butte Dam, forty miles down ditch, without any of the nasty losses caused by the river's natural channel, the LFCC was hailed in the 1950s and 1960s as the savior of New Mexico's Rio Grande Compact problems. By the 1980s, with the compact deficit made up and the LFCC in poor repair, managers decided the LFCC was no longer worth maintaining. Shortly thereafter the environmental community, pushing for the recognition of instream flow rights, came to regard the LFCC as a symbol of everything that was wrong with Rio Grande water policy. By 1994, Denise Fort, a leading environmentalist and at the time the head of the Water Resources Program at the University of New Mexico, wasted no words. "The Low Flow Conveyance Channel," she said, "is the first thing we need to get rid of."[31]

Instead, the channel morphed once again and became an unlikely but integral part of the MRGESCP.[32] A 2003 Fish and Wildlife biological opinion required the Rio Grande riverbed to be kept wet below San Acacia Diversion Dam during certain critical periods in the life of the endangered silvery minnow. But even if the MRGCD diverted no water, the river would still go dry. Where could the MRGESCP possibly find additional water to meet the ESA requirements?

Lightbulbs went on in the heads of water managers throughout the middle Rio Grande: the answer was the nearby no-longer-used LFCC, itself both a partial cause of the dry river and a source of water to keep it wet. Over the fifty years that the LFCC lay next to the Rio Grande, the relative heights of the two water channels had changed with respect to where they met near the MRGCD's San Acacia Diversion. The river had aggraded. By 2003 in the middle Rio Grande, the river bottom was in places eighteen feet higher than the LFCC west of it.

Now Steve Hansen navigated his truck along this channel, into which no water had been intentionally diverted since the 1980s. A few miles below the diversion dam, a trickle of water appeared in the bottom of the channel, a result of the rising river bed and the falling channel and the hydraulic connection between the two. Mile after mile that trickle grew, until, fifteen miles from the Elephant Butte Dam, the water in the channel flowed hard and steady. Part of the flow leaked into the channel from the higher river to its east and part of it leaked in from groundwater and return flows from the irrigated fields to the west. There it was: supplemental water, available when the endangered species needed it.

The fledgling MRGESCP had in 2004 authorized the construction of four pumping stations along the eastern banks of the LFCC at places parallel to the river, only separated by a levee and the space of one hundred yards. (The pumping stations replaced the jury-rigged pumps that Reclamation had been toying with since 2000.) As soon as the river began to dry during critical stages of the silvery minnow's life cycle (such as the breeding period after snowmelt had run its course and before the summer monsoon started), the new pumps were turned on, water was sucked out of the LFCC and pumped over the levee and fed into the dry river, keeping it wet.[33] The pumps saved the minnows for another day. Everyone once again breathed a sigh of relief.

This new solution to the silvery minnow problem was especially ironic because the efforts of the Rio Grande to reassert its natural pathway threatened to destroy the newly discovered usefulness of the LFCC. Water always seeks lower ground. Below the San Acacia Diversion, the river inched toward lower ground to the west, incising its western bank and moving toward the levee that constrained it. It was only a matter of time before the river would erode the levee away from the bottom and overrun it at the top. If that happened, the river would pour unimpeded for miles to the west until it encountered a higher elevation, almost as far as Socorro itself. The 10,000 irrigated acres between the river and the mountains to the west would become a large lake. Then the river would carve itself a new channel far west of its present location, and a new Rio Grande world would form.

To prevent this sequence of events, the MRGESCP elected in 2005 to participate in a restoration project designed to strengthen the levee and protect the lower fields to the west. Blending funds from a variety of sources, officials moved a mile and a half of the existing levee and LFCC a short distance to the west, away from the encroaching river. The new, improved levee would continue to protect the low-lying agricultural lands from the aggrading river, which otherwise would have moved much more radically to the west.[34]

Everyone breathed a sigh of relief.

While managers monitored the relationship between rising river elevations to the east and static levee heights to the west, real trouble snuck up from the south. In 2005 a sediment plug formed in the Rio Grande just downstream from the new levee, below Bosque del Apache National Wildlife Refuge, a crucial way station for migratory birds on the Rio Grande flyway. The natural plug began to back up the river as effectively as the biggest man-made dam might have done. Unless something was done quickly, the backed-up river would overtop the protecting levees and completely submerge the wildlife refuge and its unique habitat to the west.

In desperation, federal officials turned to Reclamation and comman-deered from Louisiana a huge amphibious excavator, the likes of which had served the federal government in its fruitless efforts to save New Orleans from the Mississippi. The monster lumbered onto the Rio Grande and cleared the plug twice in 2005 and 2007. Everyone breathed a sigh of relief.

Reclamation officials then acquired two of the dredging monsters for the San Acacia reach of the Rio Grande. Today, the giant amphibious excavators serve as reminders of the similarities between this reach of the Rio Grande and the lower Mississippi: in both cases an expensive and ultimately inad-equate system of levees control and constrain great American rivers that want to move somewhere else. In the case of the Rio Grande, near San Acacia, it is miles west, to the foot of the Lemitar Mountains. The environmental move-ment has not succeeded in returning us to a full respect for the river's wishes.

Reclamation's Hansen pulls no punches. "The most spectacular train wreck on the horizon," he says,

> is the impending breach of the levee system that protects the Low Flow Conveyance Channel from the river. The river channel averages fifteen feet above the valley floor. The water, with all of the force of nature, wants to go to the lowest spot. It's not a question of if, but when. That would destroy the existing silvery minnow and flycatcher habitat more surely than any human event. We did consider a project some years ago that would help the river do what it wants to do, move to the west side of the Socorro valley. But the economic costs were way too high. The political costs were impossible.[35]

The sighs of relief may have been premature.

Saving Human Systems: The Revenge of the Acequias

New Mexico's remaining community ditches, or acequias, although mar-ginalized and only minimally supported in their infrastructure needs, still maintain a fragile toehold in northern New Mexico. The late Reynolds always viewed the four hundred or so still functioning acequias as anachronisms whose use of water was barely beneficial. As long as new groundwater was available for a growing New Mexico, acequia consumption of tributary surface water didn't do much harm in the grand scheme of things. But from the mid-twentieth century on, officials recognized that groundwater pumped for new development could no longer be free. To withdraw more groundwater, new appropriators would have to retire surface water uses. As water grew scarcer at the end of the twentieth century, the ancient acequias of northern New Mexico and their water rights became a tempting target for developers of new projects for new residents.

Richard Cook of Espanola, New Mexico, led the charge. Cook's father, Wendell, had run one of the leading mercantile stores in the predominantly Hispanic town. He helped bring credit to the northern Rio Grande for the first time and also introduced a new element of financial risk. It was no accident that the Santa Cruz Irrigation District, set up in the 1930s, ended up transferring to Wendell a good deal of land within the district for which the Hispanic farmers could not pay.[36] Richard Cook was a chip off the old block. Gruff and well respected, even feared, in the community, Cook exploited the resources that he found or could get, regardless of the impacts. He gouged out hills in sight of highways for the sand and gravel they would yield. He mined pumice within sight of the Valles Caldera for material needed to stonewash jeans at a factory in El Paso.[37] When residential construction peaked in New Mexico at the turn of the millennium, he went into the subdivision business. That's where Cook smacked up against the acequias.

At one subdivision north of Espanola, county planners insisted that Cook secure water rights for the new homes before he offered the land for sale. At a second development, when he dug a pit to obtain the needed sand and gravel, lo and behold, it filled with water. For both the pond and the home construction, Cook needed to secure existing water rights, and he found them, he believed, in two tracts of land served by two ancient community ditches, the Acequia de Hernandez and the Acequia de Gavilan, north of Espanola. Cook found two beleaguered tract owners below the ditch willing to sell to him. Long-existing law established that the water rights belonged to the tract owners, not the community ditch. He thought he was in the clear.

By then, however, the acequias had been organized for the first time ever under a statewide umbrella organization. In 2003 the extremely independent acequias joined forces to push back against prior appropriation, and the acequia association slipped a bill through the state legislature giving acequias the right to approve or disapprove transfers of water rights away from their community ditches. The Acequia de Hernandez and the Acequia de Gavilan had both adopted rules requiring such prior approval. When Cook applied, the acequias turned him down, considering such a transfer would be detrimental to their communities. Cook immediately went to court.

Between 2004 and 2008, the acequia case worked its way up the judicial ladder. Finally, in late 2008 the New Mexico Court of Appeals decided that the statute authorizing acequias to decide when individual rights could or could not be moved away from the ditches that served them was indeed constitutional. The Cook case went back to the lower court for another round of new determinations on the legitimacy of the criteria that the two acequias had applied in turning Cook down.

In the meantime, acequia after acequia in northern New Mexico considered adopting rules giving themselves veto control over transfers. In January 2008 Chimayo's Acequia de la Cañada Ancha and other acequias called a meeting of their two hundred members to consider such a rule. Wracked by one of the worst drug problems in rural northern New Mexico and torn by social tensions between Anglo newcomers and ancient Hispanic families, Chimayo hardly seemed a promising place for a strong community stand on the value of water.[38] The situation was exacerbated when a new member of the community circulated inflammatory claims that the proposed restrictions on water transfers would deprive water rights owners of their last and most valuable property rights. Nevertheless, on a freezing winter night, Chimayo irrigators packed the parish hall at the local elementary school and voted 198 to 2 to adopt the antitransfer rule and restrict their own right to transfer their private water rights.

Actions such as this in the early twenty-first century represent a move away from the commodification of water in New Mexico and toward reestablishing its community value. Tensions persist between community and environmental values and the free-market dogma that prevails in American culture. Community values do make it more difficult for water rights to move via markets to their highest and best economic use, a movement essential in a world of finite water resources. Frustrated by the obstacles raised by some acequias, developers cast a wider and wider net in search of existing water rights to support their new uses, looking two hundred miles down the Rio Grande and across watersheds to find existing rights to purchase and move. And Cook? He found there were limits to what communities were willing to trade for cash and was forced to abandon his development plans, at least temporarily.

In the meantime, the acequias and other well-established New Mexico water institutions struck again, this time insulating themselves legally from the inevitable conversion of most Rio Grande surface water irrigation rights to municipal and industrial demands. In 2009 the New Mexico

State Legislature, in a long and awkward new statute, exempted acequias and conservancy and irrigation districts, among other holders of ancient rights, from the power of municipal condemnation. Cities no longer could condemn surface water rights to offset the inevitable effects of their existing and massive groundwater withdrawals. The law and hydrology finally had diverged: municipalities no longer had the ultimate legal power (condemnation) to do what hydrological reality said that they had to—account for their effects over time on the river.

Santa Ana Pueblo Leads the Way in Restoration

Myron Armijo wears many Rio Grande hats. As a native son of the middle Rio Grande Santa Ana Pueblo, he was born and raised on ancient Pueblo lands through which both the Jemez River and Rio Grande run. In 2009 he was serving as the pueblo's lieutenant governor and was employed full-time by the Office of the State Engineer, bridging worlds that have often been chasms apart. In all of those roles, Armijo was uniquely positioned to involve his pueblo in the restoration work of the MRGESCP on the Rio Grande.

Led by Armijo, Santa Ana Pueblo joined the MRGESCP in 2003 and indicated its interest in restoring the six-mile stretch of river that ran through the bordering Pueblo lands. Santa Ana Pueblo chose to focus its attentions on clearing from 1,300 acres on the west side of the river the invasive species—the saltcedars, Russian olives, and other non-native plants—that had taken over the riparian area and made of it an impenetrable jungle.[39]

The difficulty of getting permanently rid of the invaders had tormented New Mexico water managers in decades past. In the 1960s they advocated using flamethrowers. (It was a joke. The one time they tried it, they set the bosque on fire.)[40] In the 1970s herbicides sprayed from low-flying planes were the chosen poison.[41] (Borne on by the winds, the herbicides often ended up where they shouldn't have been.) In the 1980s they tried root plows. (The invasive species came right back.) And in the 1990s they tried herds of goats, imported from Oregon along with their shepherds.[42] (Thieves stole the goats. One was snatched for an animal sacrifice.) But the Bosque del Apache National Wildlife Refuge near Socorro has had some recent success in eradicating saltcedar and restoring a diverse undergrowth of native marsh plants. Similarly, when Santa Ana Pueblo's turn came in the 2000s to take a stand against the invasion, the pueblo didn't fool around.

The pueblo located a monstrous and peculiar machine in Texas, one as huge as the Louisiana amphibious excavators that Reclamation would use in 2005 to clear the sediment plug downstream of the Bosque del Apache. The machine pushed an enormous metal drum nearly the size of a house from which protruded giant solid metal spikes. As the drum was driven forward, it tore up everything in its path, above and below the earth. It swallowed the materials in its belly and disgorged the digested material as mulch in its wake.

This strategy worked. The machine left a bosque that few had ever seen, a bosque free of the jungle of species that had invaded the area as a result of years of river management. Only the cottonwoods were left standing. Walking paths now wended through a relatively open savanna from which visitors could clearly see the sparkling river (see plates 26 and 27).

Many state and federal agencies participated in the Santa Ana Pueblo restoration effort. Many different benefits ensued. To state officials interested in increasing river flows, the elimination of water-hungry phreatophytes like saltcedar held the promise of more water for the river.[43] It didn't hurt tourism, either. The new, clear bosque offered visitors a bucolic walk along a natural paradise in addition to the draw of Santa Ana's casino and resort.[44] And for the pueblo members themselves, the restoration offered a vision of the land and the river running through it that hadn't been present for decades.

When Santa Ana Pueblo lieutenant governor Armijo talks about the project these days, it is this last aspect that he most emphasizes. Tears come to his eyes as he tells of taking pueblo elders along the new paths. Tears come to their eyes, Armijo says, when they see a river world restored to what they saw in their youth.

Saltcedar: Scapegoat or Scoundrel?

Like many imported plants that prove particularly well adapted to a new environment, saltcedar's reputation has been tarnished by its success. Originally brought from Eurasia to serve as a windbreak and to control erosion, saltcedar (*Tamarix* spp., also commonly known as *tamarisk*) came to be reviled in areas of the Southwest in the mid-1900s, where it came to dominate the precious riparian corridors of New Mexico and Arizona and choke out native plants such as Rio Grande cottonwood. At the time it earned the title "Water Vampire of the West" for the large amounts of water it was supposed to suck out of the rivers.[1]

As we have seen, the plant is extremely difficult to eradicate. Its critics say it hogs the river water, reducing downstream flows through evapotranspiration (ET). It also accumulates salt and excretes it onto its leaves, which drop and leave the soil sufficiently saline to deter germination of native species. It has been considered a poor substitute for native vegetation in terms of providing habitat for riparian creatures such as beloved species of subtropical songbirds.

But does it deserve the bad reputation? Its invasive nature may in fact be a by-product of modern engineering and technology. Recent studies suggest that altered (i.e., managed) streamflow is a prime cause of increased saltcedar dominance, and that its control can only be achieved with water management regimes that at least mimic climatic norms for flood timing and intensity. The distribution of saltcedar, cottonwood, and willow in Arizona shows that rivers with natural flood regimes and perennial flows are abundantly populated with cottonwood and willow but with relatively little saltcedar, while those with dam-regulated or intermittent flows are dominated by saltcedar.[2] Decreased flows are to blame for the shift from cottonwood and willow to saltcedar, rather than saltcedar being the culprit that decreases the water.[3]

And just how much water does it consume? Old methods of measuring potential ET estimated that saltcedar consumes ten to twelve feet of water per year, but new techniques that measure plant water use over wide areas show that saltcedar actually uses an average of only 3.3 feet of water per year, comparable to what native riparian vegetation consumes.[4] In the areas of the middle Rio Grande where the water table was shallow, saltcedar stands used the same amount or less of water than cottonwood stands.[5]

As for songbirds, saltcedar has proven to be one of their favored habitats. Songbirds preferred a mix of 70 to 80 percent saltcedar and 20 to 30 percent native vegetation in one study.[6] Eradication efforts have also been stymied by legal and ornithological considerations. It turns out that saltcedar is a favored habitat of endangered Southwestern willow flycatchers, for example, a discovery that has halted eradication efforts—which have included bulldozing, herbicides, and biocontrol insects—in areas where the birds have taken up residence. Nevertheless, virtually no other fauna find life habitable beneath a thick saltcedar canopy, and so saltcedar may replace a rich and varied habitat zone with a single plant that provides no food, no grazing, and no shelter, except for birds.

Biologist Juliet Stromberg and colleagues point out that there is a substantial communication gap between science and management over saltcedar. Scientists and managers alike need to at least examine their antiexotic bias and avoid war-based and pestilence-based terminology to encourage an objective and reason-based perspective on the maligned plant.[7]

Notes
1. Hall 2002a, 60.
2. Stromberg, Beauchamp, et al. 2007; Stromberg, Lite, et al. 2007.
3. Lite and Stromberg 2005.
4. Nagler et al. 2008; cited in Glenn et al. 2009.
5. See Glenn et al. (2009) and Shafike, Bawair, and Cleverly (2007).
6. Van Riper, Paxton, and O'Brien 2008.
7. Stromberg et al. 2009.

Adjudication Nightmares on the Rio Grande

In 1996 the Houston-based consular representative from Spain offered to come to the University of New Mexico to deliver a lecture. The dean of the law school asked what he would like to talk about, and the consul, who was a history buff, said that he would talk about Queen Isabella's will of 1492. The dean, who wasn't a history buff, agreed, having little idea what the consul was talking about. The dean worried that no one would show up for a talk like that. To avoid the embarrassment of an empty hall, he scheduled the lecture in a small room.

On the evening of the lecture, the dean and the Spanish consul were shocked at the crowd milling around the entrance to an already packed room. There were so many lawyers, historians, and Native American leaders in the audience that they had to move the talk to the law school's largest lecture hall. Even in the larger venue, it was standing room only.

What could account for such widespread interest in such an esoteric and ancient topic? As it happens, Isabella's 1492 will is one of the foundational documents of Indian law, forming the basis of Pueblo claims to water arising under Pueblo, Spanish, and Mexican law, as guaranteed by the 1848 Treaty of Guadalupe Hidalgo.[45] The audience had turned out to find out what the Puebloan rights were in 1492 and still might be five hundred years later. In the late twentieth and early twenty-first centuries, the issue was crucial to the formal definition of Pueblo Indian rights in Rio Grande stream-system adjudications.

Recall that Reclamation's Morris Bien had brought the formal doctrine of prior appropriation to New Mexico and the West around 1907. His codes required the formal licensing of water rights by a state engineer to determine how to apportion limited public water supplies among individual claimants. This meant formally recording the priority, quantity, and source of water for each water right subsequently established.

However, the Bien Code also recognized the validity of water rights that predated the state-adopted codes. This was one of the reasons for *stream system adjudication*, a method by which courts would recognize, confirm, and define previous rights so that they could be incorporated into the list of newly licensed rights and administered along with them. In most western states, there were few precode uses of water, mostly dating from the 1850s. But in New Mexico, thousands of water claims originated before 1907, some stretching back to Isabella and earlier. These would have to be incorporated into one law of the Rio Grande so that all of the claims to the common river source could be put together.

Currently, there are twelve stream system adjudications pending in New Mexico, six of which focus on the Rio Grande Basin.[46] The pending Rio Grande adjudications range in size from one thousand defendants in the Santa Cruz adjudication (involving a tributary of the Rio Grande near Espanola) to more than five thousand water rights claimants in the lower Rio Grande adjudication, in New Mexico below Elephant Butte Dam. All in all there are perhaps fifteen thousand water rights claimants in these adjudications. None is yet complete and the staggering number of claims to be sorted out and interrelated explains part of the delay.

But the real breadth and depth of adjudication nightmares on New Mexico's Rio Grande are displayed in the *Aamodt* lawsuit filed in 1967. (Since the litigants were listed in alphabetical order, R. Lee Aamodt, a scientist from Los Alamos National Laboratory who had a house in Nambe, New Mexico, was in an unbeatable position for claiming the case name.) Today, the *Aamodt* case—involving twenty-five hundred claimants (including four Pueblos) to a small Rio Grande tributary stream system just north of Santa Fe—enjoys the dubious distinction of being the oldest unresolved federal lawsuit anywhere in the United States.[47] The determination of the nature and extent of Pueblo rights to a very limited supply under the laws of New Mexico's previous sovereigns certainly has slowed things up. Yet the forty-five-year-old *Aamodt* case is nothing compared to the five-hundred-year-old probate dispute over Isabella's 1492 will.

Carlos Fuentes, the Mexican intellectual and writer, has said that old Mexico can't get to its future because it is so weighed down with its incredibly rich and long past.[48] The same goes for New Mexico and its pending water adjudications. The same also goes for the middle Rio Grande, a critical reach of the river that no one has even dared try to adjudicate.

Rio Grande claims between Cochiti Dam on the north and Elephant Butte on the south involve more than half of New Mexico's present population; most of New Mexico's economic future; six of the state's nineteen pueblos; its largest and most important irrigation district; and a substantial land-based Hispanic population whose water use dates back to the late sixteenth century. Without a formal definition of the nature and extent of pre-1907 rights to water, the water rights of all are somewhat at risk. But once (if ever) begun, the adjudication of all rights to the water of the middle Rio Grande would prove the mother of all adjudications. Compared to the seemingly interminable existing adjudications, these older claims will have higher stakes, more claimants, and more exotic claims involving more interrelated groundwater development and more very old communities.

In the face of these monumental problems, most water administrators turn a blind eye to legal rights to water in the middle Rio Grande, preferring to stumble forward in the relative darkness of an undefined water regime. Others suggest alternatives to the clumsy judicial adjudication process.[49] Still others have chosen to settle ancient claims, exchanging theoretical rights for present resources.[50] The fact remains that recognized Native American pueblos such as Isleta, just south of Albuquerque, have plausible but undetermined claims to the whole river. Hispanic communities have very early claims, also undefined. The whole structure of Anglo development, shaken by environmentalism, could be upended by these ownership questions. Without a firm grasp on the nature and extent of these Puebloan and Hispanic rights, efforts to restore the river and move it into a brave new future are all built on sand.

The City Drinks New Water from the Rio Grande

On Friday, December 12, 2008, leaders of the New Mexico water community gathered near the Alameda Bridge crossing the Rio Grande in Albuquerque's North Valley. The federal, state, and local representatives had come together to celebrate the grand opening of the City of Albuquerque's Drinking Water Project. The $400 million project involved thirty-eight miles of distribution lines in the city surrounding them and a water treatment plant eight miles from where they stood. But the real heart of what some called the most significant Rio Grande development since Reynolds brashly seized control of groundwater in the Rio Grande basin in 1956 lay in front of the assembled dignitaries.

There a brand-new 620-foot-long diversion dam stretched across the river (see plate 33). Like Cochiti Dam and two MRGCD diversion dams above it and two other MRGCD diversions and Elephant Butte Dam below it, this new dam reached from levee to levee and controlled the entire river at that point. However, unlike any other previous Rio Grande dam, this one was a state-of-the-art inflatable bladder diversion structure. It had individual segments that could be raised when the river was high, laid flat when the river was low, and could target exactly whatever amount of river water was to be diverted.

A week before the opening ceremony, on December 5, the city had opened the diversion works and let the river water enter the new system for the first nonexperimental time. If it hadn't been clear before, it was now: Albuquerque had switched from an exclusive and ultimately doomed reliance on groundwater to a system that would rely on the more sustainable surface water supplies of the beleaguered Rio Grande itself.[51]

John Stomp, the battle-hardened leader of Albuquerque's water department, had claimed for years that concerns for long-term sustainability required this transition. A whole series of sophisticated, detailed studies of the Rio Grande aquifer in the 1990s had suggested that the city's groundwater resources weren't nearly as ample as once thought.[52] Further, dependence on groundwater mining, no matter the extent of the resource, wasn't sustainable in perpetuity. State Engineer Reynolds recognized as much at the start of his regime. Now, at the turn of the twenty-first century, the city agreed. It would turn for its basic supply from the groundwater that had allowed it to grow to a desert urban metropolis of 750,000 residents to a continuously self-renewing surface water supply. Opening the gates of the diversion dam honored that new commitment.

The surface water supply on which the new project would depend was not, however, the native water of the Rio Grande from which it would be drawn. Instead, it was the San Juan–Chama water imported from the neighboring Colorado River basin that had begun to flow through the Continental Divide in 1971. From the outset, Albuquerque had paid for some of that imported water. Since 1981, it had paid for one-half of the 96,000 acre-feet brought into the Rio Grande annually.[53]

But in more than twenty-five years before the Drinking Water Project officially opened, Albuquerque had consumed little of the San Juan–Chama water it had contracted for. Instead, the city pumped groundwater for its burgeoning population and leased or lent its imported San Juan–Chama water to third parties for their use. It had made its water available for recreational uses at downstream Elephant Butte Reservoir and upstream Cochiti Lake. It chipped in some of its water to make up irrigation shortages in the MRGCD. It had even provided some of its San Juan–Chama water to bail out the endangered silvery minnow, the biggest small fish on the river.[54]

In 2009 the city did an about-face and called in its water loans. With the Drinking Water Project actually online, Albuquerque would divert annually its entire 48,000 acre-foot San Juan–Chama allotment and another 48,000 acre-feet of native Rio Grande flow at the new bladder diversion dam. The Rio Grande water would return to the river at the city's downstream wastewater treatment plant once it passed through the regional water system. To water accountants, it looked as if the Rio Grande and its waters would remain whole even if at the Alameda diversion the river was losing 96,000 acre-feet a year that it had never lost before.

It was the Colorado River, not the Rio Grande, that was providing the city's self-sustaining supply.[55] The Rio Grande was just the transport vehicle to carry the imported water to Albuquerque. The city already had outgrown the surface water and groundwater water supplies of its own watershed. In shifting to Colorado River water, Albuquerque joined the company of Las Vegas, Phoenix, Tucson, and the most rapacious Colorado River user, Los

Angeles. How, some asked, could Albuquerque choose to abandon its own watershed in the name of sustainability? Clearly, by switching from Rio Grande water to Colorado River water, Albuquerque was buying into all the problems of that basin, which was fraught with even more uncertainties than the Rio Grande.

The switch raised questions about the Rio Grande that Albuquerque was trying to leave behind. The city's casual treatment of its San Juan–Chama entitlements prior to its new total consumption of them must have influenced Rio Grande flows. The city's previous reliance on groundwater pumping had contributed groundwater to the river in the form of return flows. Both the loss of San Juan–Chama water and the loss of municipal groundwater return flows would surely alter flows into Elephant Butte and cause compact problems. From that perspective, the city's new surface-water Drinking Water Project would simply trade a new set of artificial inputs to and outputs from the river for an old set.

To his credit, in issuing the City of Albuquerque permit, State Engineer John D'Antonio did his best to minimize the risks of the leap into the unknown that the Drinking Water Project represented.[56] It was a measure of the city's desperation, or possibly arrogance, that it jumped into construction of the multimillion dollar project before the appeals process had run its course. It remains to be seen how the Rio Grande will react to the new set of engineering factors imposed on a previously engineered river.

Computers to the Rescue of a Natural River

Since the death of Reynolds, a wide array of new values has emerged that were long suppressed by New Mexico's devotion to its compact obligations and to a definition of beneficial use that was restricted to human and economic applications. Vestiges of an older, more natural Rio Grande are reappearing. Thousands of jetty jacks have been jerked out in an effort to destabilize the banks of the river.[57] Dam releases from Cochiti are now sometimes done in pulses, rather than continuous flow, in order to erode banks and sandbars and overflow the banks. Bulldozers have gouged new artificial channels and depressions to encourage the river to vary its course and create wetlands. Wetlands that disappeared seventy-five years ago have been reconstructed and now host native vegetation and wildlife. In large areas, the saltcedar and Russian olive have been thinned, and the native bosque forest is beginning to thrive once again.[58]

Real restoration has not been achieved, however. Thousands of acres of riparian habitat remain under dense stands of nonnative shrubs. Efforts to restore the links between in-channel and overbank zones, using high flows to reroute the river, are greatly hindered by houses that have been built on the floodplain, restricting the high flows to fairly pitiful levels. Along most stretches of the river the riparian habitat is confined by levees to a narrow strip along its banks. Much work remains.

To accomplish the task, river managers have embraced new kinds of technological solutions, such as careful accounting of river flows through computer modeling. For example, following the massive silvery minnow die-off of 1996, and years of court cases and negotiations, another weapon in the save-the-minnow arsenal was forged. This time the target was water stored behind federal Rio Grande dams. In 2005 a complex agreement was hammered out between environmental organizations, the City of Albuquerque, the Corps, the state government, and Reclamation.[59] Trade-offs and leases involving San Juan–Chama water imported from Colorado and native Rio Grande water were arranged and approved, all with the aim of keeping the Rio Grande wet through minnow breeding season and allowing at least intermittent flows after that.[60]

Now a schedule of water releases has been worked out, based on daily snowpack data from automated monitoring sites operated by the federal Natural Resources Conservation Service. The data include historical records of runoff and water demand and long-term weather forecasts from the National Oceanic and Atmospheric Administration that are based on a global network of weather and ocean temperature sensors. All of this information is fed into an immense computer model known by an acronym suitable for a giant worm in Dungeons and Dragons: URGWOM (Upper Rio Grande Water Operations Model).[61] The model calculates the likely snowpack in the late spring; the rate of spring warming; the amount of the snowmelt expected to actually run off into the Rio Grande; the amount of water that the agricultural fields would need to consume; the amount of water that would be lost from the river as it flows southward due to evaporation and use by streamside plants; the amount that would be lost by seepage through the riverbed; and many, many other factors. The result is the schedule of releases, a table estimating how much water will enter each reservoir on each day and how much should be released under the dam to supply the users downstream.

But estimations are only a guide to predicting actual runoff. Actual scenarios go something like this: A heat wave settles over the basin and the farmers call for much more water than has been predicted. More water evaporates from the river than has been calculated. The river water heats up more than the model has expected and consequently seeps faster into the riverbed. The end result can be a trickle of flow at a site like San Acacia, north of

Socorro, instead of the ample flows predicted. The seasonal plan often has to be adjusted on a day-to-day basis. The rules are complex.

Some water can be used for one purpose but not another, some has to be released according to set schedules, and other water can be flexibly released. The reservoirs must be operated prudently to maintain adequate reserves until the end of the season, while not kept so full that they will become uncontrolled in the event of an unexpected flood. All entities maintain their own balance sheets of water debts and water credits against their legal allotments. The waters of the Rio Grande can be swapped, traded, lent, or purchased, but the balance sheets have to close and every drop of water has to be measured and accounted for.[62]

To the extent that the river has become a tightly controlled device for delivering water where and when needed, Reynolds's vision has been achieved. He could not, however, have envisioned the extent to which these flows would be manipulated to benefit the natural system, transcending engineering designed to merely meet legal obligations bearing on the needs of farmers. Water management has adopted a broader view of beneficial use and embraced the value of cooperation among stakeholders with diverse interests. However, active water management remains more focused on responding to crises rather than addressing the ultimate sustainability of the riverine system.

Chapter X

Fulfilling Rio Grande Demands
What Has to Give?

What does the future hold for the Great River? In an age of steadily increasing pressure from growing population and technology can the river maintain any of its primordial character? In the post-Reynolds Rio Grande, the priority assigned to efficient economic uses of the river's water continues to be weighed against the importance of traditional uses and nonhuman uses. Invasions of nonnative species in the riparian corridor, increased municipal water demands, and ancient users reasserting control over their traditional rights further complicate an administrative framework that appeared much simpler thirty years earlier. And the legal structure for water rights priorities may still be turned upside down by the Endangered Species Act.

Future problems cannot all be foreseen, as Steve Reynolds could attest were he alive today. But the known challenges now facing the river are themselves formidable. The southwestern United States is predicted to be one of the prime areas to feel the effects of anthropogenic global warming. Meanwhile the population will continue to increase. The prospect of acquiring new water to meet the increasing demands is dim. The turn of the century provided a harbinger of things to come. In 2001, for the first time, the Rio Grande failed to reach the Gulf of Mexico, stalled by a sandbar that formed three hundred feet from its destination.[1] Since then, the area has been dredged and re-dredged, but the remaining trickle of river rarely reaches the gulf. And yet, the opposite scenario has also occurred. Severe flooding hit the Rio Grande in south Texas and northeastern Mexico in July 2010 following a hurricane and a tropical depression whose precipitation simultaneously overwhelmed three of the four basins contributing to the lower Rio Grande. The situation—called "unique" and "extraordinary" by water managers—required reservoir releases that created a new lake even larger than the 84,000-acre Falcon Lake toward which the floodwaters were traveling.[2] Both worldwide and regional forces are acting in concert to diminish and unsettle the river.

Inevitable Climate Change

Scientists on the United Nations Intergovernmental Panel on Climate Change have agreed that continued warming of the global climate system is "unequivocal."[3] Within North America, the southwestern United States and northern Mexico will be the persistent hotspots for future climate change. As precipitation decreases, soil moisture will dry out and the occurrence of drought will increase.[4] The consequences will be felt in both the Rio Grande and Colorado River systems.

Long-term drought is no stranger to the Southwest, as paleoclimate records show. Drought has decimated landscapes, dried up inland seas, and forced out whole populations of people, plants, and animals. Droughts in the second half of the twentieth century were bad, but not as bad as some experienced in the deeper past.[5] Scientists have reconstructed annual precipitation in the Southwest for the last two millennia, looking, for example, at rings of ancient trees in El Malpais National Monument in Grants, New Mexico, and the Sandia and Magdalena mountain ranges of the middle Rio Grande basin. The 1950s drought in New Mexico is considered the third worst since AD 622, but was surpassed by much longer droughts: one lasting twenty-three years in the 1500s and another lasting twenty-six years in the 1200s,[6] the latter being a prime reason that Puebloan groups abandoned the mountainous north and eventually settled along the Rio Grande.

Climate change, drought, and extreme weather—shifts from extremely dry to extremely wet years and floods—played a big role in the history of the Hohokam and Anasazi, the most advanced ancient Southwestern civilizations, and led to their dissolution, migration, and transformation (see plate 28). With exponentially larger populations in the present-day Southwest demanding much more water for industry, crops, and personal use, and the inevitability of climate change, we must learn from their experience, acknowledge the potential for disaster, and prepare to adapt to new realities.

Climate change will have enormous impacts on water resources of the Rio Grande basin. Changes in runoff, streamflow, the amount and timing of snowfall, snowpack, and snowmelt have already made water planning a risky business. In announcing an "end to stationarity," several distinguished climate change and environmental scientists have suggested that water managers can no longer assume that natural systems will work within constant and defined boundaries of variability.[7] (*Stationarity* is a statistical term meaning that typical values—for example, streamflow averaged over thirty years—remain the same regardless of what thirty-year period is picked.) The boundary of what is normal has disappeared. Our historic data are, at best, less predictive of the future; at worst, they are dangerously misleading. The future will be different from the past. The following changes are predicted for the Rio Grande basin:

Increased aridity, more severe droughts

Severely dry conditions by 2060 may well be the norm rather than an unfortunate unusual event. Drought used to be driven by precipitation, but perpetually higher temperatures alone will cause enough drying to create a near-constant state of drought.

Scientists who monitor droughts often use an index called the Palmer Drought Severity Index to examine the combined effects of precipitation and temperature on soil moisture. Simulations of this index in a warmer future climate indicate that soil moisture in the Four Corners area by 2050 will average the conditions experienced during the most severe historical droughts (see plate 29).

Changes in snow, snowpack, and runoff

Snowpack in the San Juan Mountains efficiently stores water that eventually melts and flows into and down the Rio Grande in the late spring when thirsty young crops are emerging, clouds are scarce, and rain rarer still. As the planet warms and the balance of rain and snow shifts, soil moisture, runoff, streamflow, and groundwater recharge will change.[8] There will be comparatively less snow and more rain, with a runoff peak too early in the spring to be used for irrigation. As precipitation diminishes, the amount of runoff also decreases in a nonlinear fashion. For example, if precipitation decreases by 10 percent, runoff will decrease by more than 10 percent.[9] Arid and semi-arid regions like the Rio Grande basin become more vulnerable to droughts, floods, and crop failure.[10]

More extreme events

With the shift from snow to rain, more winter flooding can be expected throughout river basins in the Southwest. Heat waves are also expected to increase in number, intensity, and duration, affecting crops as well as human health.[11] A warmer world also causes increased evaporation and more intense precipitation.

Transformative landscape change

As we have seen, past changes to the landscape from logging and cattle grazing greatly affected river flows, the river course, water quality, and soil alkalinity. The Rio Grande landscape is undergoing massive new changes that are the indirect results of human-caused climate change.

Pine forests have recently experienced massive die-offs from bark beetle infestations triggered by drought-induced water stress throughout the Four Corners states, particularly in New Mexico. An aerial study showed 67 percent mortality of piñon pine in Cochiti Pueblo in 2002, a loss of 4 million trees. [12] A 2006 survey of the Mesita del Buey (near Los Alamos) in northern New Mexico showed piñon mortality of 90 percent.[13] Insect damage, drought-induced mortality, and a prolonged fire season have also made forests vast acres of tinder; the West has already seen a six or seven-fold increase in fires in the twenty-first century, compared to the early decades of the 1900s.[14] All of these changes in the uplands will affect the rivers that are at their heart, changing both the amount of water in the rivers and the timing of the water's arrival.

Water Demand

Population growth has slowed in the Rio Grande valley in recent years, but continues, especially in the urban areas, at a rate much higher than in most of the United States. In 1940 Albuquerque was the only metropolis in the entire Rio Grande valley, but it claimed this status with a population of merely 35,000 people, a small town by national standards. During just the next twenty years, the population surged to more than 200,000. Each person used about 200 gallons of water every day, which meant that the water consumption of the city zoomed upward by 33 million gallons per day. Where did that water come from when the Rio Grande surface flows were fully appropriated? Albuquerque couldn't just build a dam across the Rio Grande and begin diverting its waters; somebody downstream had a claim on every drop. The answer, as we have seen, was that the city fathers frantically drilled wells to tap into the aquifer below, without regard for the consequences. Ultimately, Reynolds let groundwater pumpers have free rights to groundwater until the slow but inevitable effects would be felt on the already fully appropriated Rio Grande. This abundance of water, free for the taking, helped fuel Albuquerque's explosive growth spurt in the 1950s and 1960s.

After the 1960s, the rate of growth slowed but certainly didn't stop. By the 1980s, Bernalillo County (which includes many of Albuquerque's outlying suburbs as well as the city itself) contained more than 400,000 people, and by 2000 there were approximately 550,000. That meant the new population required 110 million gallons per day more than in the 1960s. Where did it come from? The City of Albuquerque bought some farms on the Rio

Grande and took them out of irrigation to try and mitigate the effects of groundwater pumping that had begun to impact the river. But the main fix was the San Juan–Chama Project, the transbasin diversion of water that by 1970 was supplementing what the Rio Grande could provide. By importing water from the Colorado River, Albuquerque snuck around the zero-sum limitation of the water balance, but the trick cannot be repeated.

After all of these adjustments, populations have continued to increase to the tune of 7,000 new residents in Bernalillo County each year for the last thirty years. If that steady pace continues, by 2050 there will be almost 1 million people in Bernalillo County and 50 million more gallons of water than in 2005 will be required each day, or about 56,000 acre-feet per year. (To put this number in perspective, under the Rio Grande Compact most of the state of New Mexico, between Otowi and Elephant Butte, is allowed to only consume up to 405,000 acre-feet of water per year.) Of course, not all of this water will be lost; typically half of it returns to the river as treated wastewater.[15] But the greater Albuquerque area already consumes about 60,000 acre-feet per year and the projection is that another 28,000 acre-feet per year will be needed.

To sum up, Albuquerque has twice side-stepped the limitations of a finite water supply: first by temporarily exploiting nonrenewable and already committed groundwater and then by exerting the political leverage to import water from the west side of the Continental Divide. Neither of these fixes is likely to be repeatable. If the metropolis is to continue to grow, it must identify specifically where the water would come from in a system where some person or entity already owns every drop of available water.

Nor are Albuquerque and its sister city across the Rio Grande, Rio Rancho, the only growing cities on the Rio Grande. From 2000 to 2008, the population of El Paso, Texas, increased 10 percent, from 680,000 to 750,000.[16] Continued growth at this pace would mean it will top 1 million by 2036. Just on the other side of the Rio Grande in Mexico, Ciudad Juarez is growing even faster. The 2008 population was about 1.6 million and is increasing by about 13 percent per year.[17] At this rate, which is surely not sustainable, Ciudad Juarez will have a population of more than 7 million by 2036. Every person added consumes water every day.

Water not only slakes the thirst of growing populations, it drives the turbines that run the power plants that keep the lights on and the groundwater pumps sucking throughout New Mexico. Vast amounts of water are used to produce energy. Conversely, we use energy at a high rate to pump water from great depths, pipe it around our localities, treat it, deliver it to where it is used, recollect and retreat it, and dispense with it. As populations increase, energy demands rise along with water demands. The current water-energy relationships are intensifying because new sources of water are more distant, deeper, or polluted/brackish/salty, so they require more energy to purify and deliver. Some new sources of energy—nuclear, solar thermal, oil shale, fossil-fuel burning coupled with subsurface carbon sequestration, and biofuels—reduce carbon dioxide emissions significantly, but also increase water consumption, particularly in semiarid settings.

Where will all this new water come from?

Water Supply

Where does the population get new water? One approach to addressing the dilemma of water scarcity in the face of growing demand has been to search for additional sources in the Rio Grande basin and beyond. Seekers have cast a wide net in recent years.

A headline in the November 1, 2008, issue of the *Albuquerque Journal* proclaimed "Sandoval Aquifer Tests Show Huge Supply of Briny Water." Under the heading, the article quoted the county commissioner Jack Thomas rejoicing, "The implications for Sandoval County are huge."[18] This reaction was remarkable. Just a few years ago, a deep test well that produced only brine would have been considered an expensive failure and a major disappointment. Why was it now a huge success?

Desalination

Part of the answer lies in a reassessment of the geology. Based on hydrogeologic studies conducted in the late 1950s, Albuquerque was thought to sit on top of a gigantic bowl composed of clean sand and filled with fresh water. One of the earliest skeptics was John Hawley, a curmudgeonly geologist who had worked for decades in the New Mexico Bureau of Geology and Mineral Resources. By the late 1980s and early 1990s, Hawley increasingly realized that the idealized model of the basin in vogue in the 1950s and 1960s did not correspond to the sediments he saw actually coming up the hole when new wells were drilled. In 1992 he and his colleague Steve Haase put out a report that laid the hydrogeological evidence on the table.[19] The "Lake Erie" of fresh water under the city was a desert mirage, revealed now to be a briny sea. The coarse sands filled with fresh, pure water were just a ribbon running down the middle of the basin; most of the rest had abundant fine sediment mixed in, and the water in those fine sediments was often salty.

The time of pumping groundwater on easy credit is over; the hydrogeological recession has set in. Continuing to pump groundwater even far from the river is now not much different from sucking the water straight from the river itself. The anticipated easy solution, drilling deeper to tap the under-ground Lake Erie, was spoiled by the gloomy realism of Hawley and his col-leagues. In response to the tightened future groundwater supply, the state engineer has begun to crack down on new applications for large pumping projects, requiring that surface-water rights to make up the river depletions had to be purchased before the effects of pumping reached the river.[20]

Nowadays, against a background of suddenly plummeting groundwa-ter supplies, the deep aquifer filled with salty water under Sandoval County begins to look like a godsend. In 2008 Rob Sengebush, the drilling-project manager for consulting company Intera stated that the deep, salty aquifer was isolated from the Rio Grande aquifer by a fault to the south and a layer of clay-rich rock above.[21] Thus, in theory, the brine aquifer is like an underground tank, sealed off from the Rio Grande hydrologic system above it and available for unlimited pumping. Of course, it would have to be desalinated, but that technology is already well advanced. County officials announced that they had enough water to supply a city of 300,000 people for one hundred years.

The Sandoval County drilling success sparked a "brine rush" of claims to the underground salty fluid that culminated with a Canadian company, Lion's Gate Water LLC, filing a notice of intent to appropriate all the salty water under the entire state of New Mexico.[22] The rush to lay hands on all that brine grew so frenzied that on February 15, 2009, the *Albuquerque Journal* editorialized "Umpire Needed in Race to Mine Brackish Water." By the end of March, the state legislature was forced to pass a bill granting the state engineer authority over the drilling of deep brackish water wells (2,500 feet or more below the land surface). Between the time when the House of Representatives approved the bill and it was approved by the State Senate, 600,000 acre-feet per year was legally claimed in this latest western liquid gold rush.[23]

Can brine really solve the water supply limitation of the Rio Grande basin? Desalinization technology is certainly mature. Large populations, such as on the Arabian Peninsula and in Israel, depend on it for their everyday water supply. El Paso has already jumped on board and has constructed the world's largest inland desalination plant that is producing a quarter of the water it needs each day—about 30 million gallons.[24] Removing the salt adds to the cost of the water, but not prohibitively. Standard treatment of surface water (to sterilize pathogenic organisms, remove silt, and so on) costs between $1 and $2 per 1,000 gallons. Desalination only increases that cost to $3 to $5.[25]

Perhaps more important is the underlying issue of the energy required in desalination. To produce the electricity necessary to make drinkable a million gallons of water from the Rio Grande, an electric power station has to burn about 180 pounds of coal. To desalinate the same amount of brine under Sandoval County would require burning about 2,680 pounds.[26] The increased coal production and combustion would incur additional environmental costs in terms of damage to the landscape, water needed for processing, and carbon dioxide added to the atmosphere.

Bruce Thomson, Regents' Professor of Civil Engineering at the University of New Mexico, has made another important point about saline water as a potential new water supply. He writes, "We must recognize that it's not 'new' water, it's old water. Very old."[27] He is referring to the fact that this water entered the underground formations millions of years ago. Unlike ordinary groundwater, which is continually recharged (although at a very slow rate in most of the Rio Grande basin), it is a relic of the geological past. Once pumped out, it is gone. Thomson questions the wisdom of building entire cities that are dependent on a finite and, even relative to the time scale of human history, rather small supply of fossil water. He suggests that instead it be reserved for emergency use during times of drought.

There is a final, very troubling issue with desalination. Exploiting the saline water resource may endanger the supplies of active, fresh groundwater upon which we depend. Heavy pumping of these deep saline aquifers, about a mile underground, under the assumption that they are fully isolated from the active groundwater system is risky. No rocks, except possibly rock salt, are impervious to the flow of water. If a deep aquifer is depressurized by removing the water, the reduction in pressure will slowly but surely work its way upward to the surface, where it will pull down the water table. The key is being able to quantify how slow "slow" really is. Depending on certain aspects of the hydrogeology, impacts could be felt in decades instead of millennia. Near the Rio Grande, this will increase the rate at which river water leaks down into the groundwater system. Exploitation of this deep water resource could ultimately suck dry the same Rio Grande it was intended to protect. Responsible exploitation of deep saline groundwater is not impossible, but should be approached conservatively and with great caution. Unfortunately, such cautious, conservative development is unlikely in a gold-rush-like mentality of "if we don't drill and pump it right away, somebody else will."

Salt and the Rio Grande

Where it spills out of the San Juan Mountains into the San Luis basin, the Rio Grande is a sparkling, fresh river. Its water more than suffices for practically any human use. It contains 40 milligrams of dissolved solids per liter of water (a measure of the total amount of salt and other dissolved inorganic matter it contains), making it only a little more salty than water sold in supermarkets as "distilled." When the water arrives at Albuquerque, it is still good quality but contains on average about 220 milligrams dissolved solids per liter.[1] This is enough to give it a slight taste and to cause mild problems with industrial uses. Water leaving Elephant Butte Reservoir contains double this amount, and by the time it gets to El Paso, the dissolved solids are about 700 milligrams per liter. This is well above the EPA recommended drinking water limit of 500 milligrams per liter and causes a distinctly salty taste and major problems for scaling of pipes and for industrial use.[2] Finally, where the river peters out near Fort Quitman, Texas, the dissolved solids are normally well above 1,500 milligrams per liter. The main use of the water is for agriculture, but only unusually salt-tolerant crops such as cotton can survive irrigation with water so salty.[3] No one would willingly drink it if an alternative were available.

What causes such an enormous increase in the saltiness of the Rio Grande? The problem has been recognized for more than one hundred years. Agricultural engineers of the early twentieth century blamed it on irrigation. They said the salt was due to the water's "use and reuse" in multiple irrigation districts.[4] In other words, salts were being left behind from evaporation as the water was repeatedly diverted from the river, returned by drains, then diverted again downstream. Later researchers recognized that dissolved solids were actually originating from within the basin and were not simply concentrated by evaporation, but the source remained mysterious.[5]

Research funded by the Science and Technology Center for Sustainability of semi-Arid Hydrology and Riparian Areas (SAHRA) began in 2000 to take a new approach to the problem. Rather than just tracking the concentrations of major dissolved solids, researchers fingerprinted the sources of salt using isotopes and rare elements. Results clearly showed that much of the increase in salt as the Rio Grande flows downstream comes from deep brines that slowly leak into the river by moving upward along geologic faults.[6] The mixture of river water and brine is further concentrated by evaporation during irrigation, but without the brine leakage the dissolved solids would be much lower than they are.

Knowing this means that policies that reduce the amount of agriculture will not significantly improve the water quality of the Rio Grande. The locations where the brines are entering the system appear to be separate and distinct, and so a more effective approach might be to drill down and intercept the brines before they reach the river.[7] In any case, mitigating the downstream salinization of the Rio Grande requires a solid understanding of the sources of the salt and how it moves through the system.

Notes
1. Bastien 2009, appendix A.
2. U.S. Environmental Protection Agency, "National Secondary Drinking Water Regulations," 40CFR143.3, http://www.access.gpo.gov/nara/cfr/ waisidx_02/40cfr143_02.html (accessed September 20, 2010).
3. Matthess 1982, 343–49.
4. Lippincott 1939.
5. Van Denburgh and Feth 1965.
6. F. Phillips et al. 2003; Hogan et al. 2007.
7. Mills 2004, 168–201; Moore et al. 2008.

Importing water

Water supplies in the Rio Grande basin have already been augmented by the San Juan–Chama diversion, which pipes water from the headwaters of the Colorado River into the headwaters of the Rio Grande. The chances of similar rescues in the future are remote. Public perception of the value of water and of the ecological consequences of massive diversions is far greater now than in 1964, when the mother-of-all water supply solutions, otherwise known as the North American Water and Power Alliance (NAWAPA) plan, was proposed.[28] Conceived by the U.S. Army Corps of Engineers, NAWAPA was essentially a plan to replumb North America. Giant dams would be built in Alaska and the Canadian Arctic that would reroute the northern rivers into an enormous reservoir in the Rocky Mountain Trench, just north of the U.S. border. From there, part of the water would flow to the Great Lakes and part would run southward throughout the western United States and then be pumped into another huge reservoir on the top of the Colorado Rockies. And from there, the water would flow southward to finally be dumped into the headwaters of the Colorado River and Rio Grande.

Had a river of pure arctic runoff actually been pumped into the Rio Grande, New Mexico would not have to live within its water means, and there would be no search for "new" water. In reality, the days when the government could fund grandiose water projects and the public would fall into line have passed. Urgent concerns were raised about the environmental impacts of so vast a reorganization of the North American water cycle. The Canadians decided that they did not want to see a large part of their nation's runoff disappearing across the U.S. border, and when the North American Free Trade Agreement was finally passed, water was specifically excluded.

Public reaction to another relatively tiny Rio Grande import scheme illustrates the current political climate in New Mexico with respect to importing water. In early 2005 Sierra Waterworks Company proposed to drill deep wells into a saline aquifer under the Estancia Basin southeast of Albuquerque and pump out 7,200 acre-feet per year of briny water. The saline water would be desalinated and piped to Santa Fe for municipal use. The idea of exporting their water created such vociferous opposition by residents of the area that within a month the proposal was disowned by the politicians who had originally suggested it.[29]

Ultimate Limits to Population Growth

The idea of limiting population growth is difficult to implement and strongly opposed by many sectors of the economy. So, let's ask the question: if water supply constitutes the final limit to population increase, how many people could the Rio Grande support? Let's narrow the question to the Rio Grande basin between the Colorado state line and Elephant Butte Dam, since the constraints of the Rio Grande Compact provide definitive limits to water supply there.

The answer depends on how much water a person needs. In 2009 the average per-capita water consumption in Albuquerque was about 160 gallons per person per day.[1] This is significantly lower than in most American cities, but far above what most people in the world live on. Peter Gleick, a noted authority on world water use, attempted to arrive at a recommendation for the minimum basic water requirement for human life. The number he came up with was 13 gallons per person per day.[2] This allowed 1.3 gallons for drinking, 4 gallons for disposal of human waste and sanitation, 4 gallons for bathing, and 4 gallons for food preparation. This is less than one-tenth the per-capita consumption in Albuquerque, but nevertheless a significant proportion of the world population lives with less than even the minimal 13 gallons per day. In at least fifty-five countries, the average daily water use is less than this minimum. In countries such as Sri Lanka, Nepal, Bangladesh, and large swaths of northern India people get by on 4 to 6 gallons per day.[3] In Somalia, Ethiopia, Mozambique, and Uganda they eke out an existence on 2 to 3 gallons per day. In these countries, most people obtain their daily water (usually contaminated) from a vendor at exorbitant prices—often a significant portion of their daily income.

Let's be generous (others do with less) and allow each person in the Rio Grande 5 gallons per day. Over the life of the Rio Grande Compact, the middle Rio Grande's authorized depletions (the allocated share of its main stem flow, plus tributaries, plus San Juan–Chama Diversion water) have averaged about 350,000 acre-feet per year. At the living standard of much of the rest of the world, this supply could sustain more than 60 million residents. Of course, the standard of living would be drastically different from the present. These residents would be desperately impoverished in comparison. There would be no running water, no sewers, no bathing. When there was a dry year and the water supply went down, millions would die of thirst.

The scenario is both absurd and horrifying. The point is this: more is not necessarily always better. There must be limits. The Rio Grande is very unlikely in the foreseeable future to return to a pristine condition virtually without human influence, as it was one thousand or so years ago. It is also unlikely to be converted to a vast slum of 60 million people subsisting on a few gallons per day. Where, in between these extremes, do we want to stop?

Notes
1. Albuquerque Bernalillo County Water Utility Authority, http://www.abcwua.org (accessed July 2009). Note that in several Native American populations of the Southwest, particularly where running water is lacking, the gpcd is dramatically lower than in urban regions. For example, for Navajos with running water, estimated usage rates are 75 to 100 gpcd, but for those without plumbing, estimated usage is 10 to 15 gpcd, and Hopi tribal members who lack running water are estimated to use only 10 to 35 gcpd (U.S. Bureau of Reclamation 2006, 14, 16).
2. Gleick 1996.
3. Ibid., table 10.

Making the Most of What We Have

The options for increasing the actual supply of water are limited. Fortunately, the other end of the water balance equation—demand—is more flexible. Several options are available for more efficient use of available resources. Water observers categorize the two most popular options as stretching existing water through conservation or switching uses through markets.

Conserve it

Water conservation education through public outreach programs, including schools, continues to be an effective means of reducing residential water demand. Santa Fe's Sangre de Cristo Water Division has an active, aggressive, and very successful water conservation campaign. Supplemented by price incentives and use restrictions, the campaign was responsible for a 40 percent decrease in water use by its customers from 1995 to 1997, from 168 to 101 gallons per capita per day.[30]

The most effective water conservation programs target what water managers call *consumptive use*, or uses that permanently remove water from local or surface storage through transpiration or evaporation. (This contrasts with *nonconsumptive use*, usually indoor water use where water is returned through drains to the sewers, where it can be treated, recharged to the aquifer or river, and eventually reused.) While important in residential and urban areas, reduction of consumptive use is potentially even more important in agricultural areas of New Mexico, which consume nearly 80 percent of the state's water.[31] Improved irrigation efficiency has the potential to make dramatic reductions in water demand in the state's farming regions. Large commercial agricultural operations can benefit most obviously from new irrigation technologies such as microsprinkler irrigation (which uses many, low-volume sprinklers close to the ground) and drip irrigation (which applies water directly to the root area of crops), significantly reducing runoff and evaporation losses as well as the costs of pumping water.

Although increased irrigation efficiency is one of the most widely promoted solutions to increased water demand, its actual effects on the water balance are complex and even paradoxical. The cost to farmers of delivered irrigation water is minimal, meaning that there is little benefit from installing expensive water-conservation technology in terms of reducing costs. The only real incentive to farmers for increasing efficiency is to use what water is saved to irrigate additional land and thus increase income. For example, a study of the use of subsidies for adopting commercial drip irrigation in the Elephant Butte Irrigation District showed that farmers given subsidies responded by increasing their total acreage.[32] These farmers were diverting the same amount of water as they did before, but using it to grow more crops. The loss of water through evapotranspiration greatly increased. The result was less water available for others downstream and for nonagricultural uses. For this reason, the New Mexico state engineer attempts to regulate actual water consumption rather than just the amount diverted.[33]

Improving water conveyance by such methods as lining canals is also frequently urged, but often it is equally chimerical as a solution to increased water demands. The irrigation water that seeps through the bottoms of dirt canals promotes a high water table and thus abundant riparian vegetation. Most of the seepage water returns to the Rio Grande through drains. When canals are lined, water consumption is reduced, but at the expense of seeps, springs, and bosque wetlands that are critical to wildlife. Seeps and springs are also often culturally and spiritually important to native peoples. Money spent on canal lining may well achieve increased crop production, but quite possibly at significant environmental cost.

Allocate it differently

At present, the large majority of water used in the Rio Grande basin goes to irrigated agriculture, but there is a steady shift to municipal uses. Continued evaluation of the appropriate proportion of available water to devote to various uses—agricultural, municipal, industrial, and residential—might help us to achieve long-term sustainability of the resource. One suggestion might be that domestic and urban water use should have precedence over agricultural use, although no such distinction is recognized in the beneficial use doctrine.

Such a policy—giving preference to urban uses—would run counter to trends elsewhere in the United States, where many local governments have enacted greenbelt laws to preserve farmland, forests, and open space. Many residents of the Rio Grande valley feel that open farmland is inherently valuable and that agricultural lifeways are worth preserving. Retained farmland acts as a safety valve for the water supply in times of drought, since farmland can be temporarily retired with consequences less severe than cutting off urban users.[34] Irrigated farmland also has a cooling effect on the ambient temperature; farms in the middle Rio Grande could actually help to cool an ever-warmer Albuquerque.[35] A certain amount of agricultural land undoubtedly provides a societal benefit. Perhaps this should be explicitly recognized so that any additional urban growth would have to take place within the constraints of that water use.

A more radical approach would be to allocate water according to the type of farming being done. For example, by replacing all open-field alfalfa production in New Mexico with hydroponic forage greenhouses, the equivalent amount of livestock forage could be grown on less than 1,000 acres as compared with the current 260,000 acres devoted to it, and with water use reduced from 800,000 acre-feet to 11,000 acre-feet.[36] Although this strategy would maintain forage production, the aesthetic and ecological benefits of open farmland would be lost.

States like New Mexico have always left the choice of beneficial uses to the holder of water rights, treating those rights in water as equally inviolate as rights to land. But the time may arrive when the public, asserting its special interest in water and using its power to define beneficial use, might wish to specify what types of crops, for what consumers, are suitable for growing in a particular climate, taking into account the water required to grow them and the energy required to ship them.

Create a market for it

Beyond whatever water can be harvested through conservation or by discovering genuinely new sources, water for new uses in the fully appropriated Rio Grande basin must come from the transfer of existing water rights.

New Mexico water law locks private water rights into a specified amount for a designated use at a particular place without regard to the relative economic value of the use so long as it is beneficial, as broadly defined. A chile farmer's water right is as secure as that of a computer chip manufacturer; if the chip manufacturer needs a new water right, he may have to buy it from the chile farmer. Presumably, the value of water to the chip manufacturer would be so much greater than the value to the chile farmer that the chip manufacturer could offer the farmer a unit price for his water that the farmer couldn't refuse. Water markets would be the medium by which a scarce and fixed amount of Rio Grande basin water would flow toward the money.

Although populations grow, the economy changes, and environmental needs for water continue to rise to the fore, water rights structures are relatively insensitive to pressure for change in the uses of water. Conservative users with ancient rights along northern New Mexico acequias are using essentially free water to grow their bean or chile crops, water that a semi-conductor manufacturer might be willing to pay hundreds of thousands of dollars for in order to increase manufacturing capacity. The insensitivity to pressure for change creates stresses for the mobile sector of society that would like to put the water to new uses. Another segment of the population sees the insulated nature of water rights as a protection for traditional lifeways and as a buffer against conversion of the rural landscape into an endless low-density suburb. Depending on one's perspective, water markets are either the path to a smooth transition to a modern world or the death knell for rural culture and landscapes.

Because of Reynolds's strategy of allowing groundwater to be borrowed from the basin water budget, the middle Rio Grande basin has in fact been committed to water market transfers from agriculture to municipal uses for more than fifty years. Water-hungry cities like Santa Fe, Albuquerque, and Rio Rancho have been allowed to pump basin groundwater on the condition that they would purchase and retire enough surface water rights to offset the pumping effects on the river when those effects reach the river. These effects are just now being slowly felt and will grow. The debt is now coming due and getting larger. To offset the total debt of existing wells to the river will require in the end the purchase and drying up of at least 30,000 acres of presently irrigated land, perhaps as much as half the acreage that is presently irrigated in the Middle Rio Grande Conservancy District.

So far, the middle Rio Grande market for water rights has stumbled along. Cities and developers have generally been able to find and transfer the water rights their projects have required. But there have been glitches. Existing water rights available for market transfers are ill-defined. Like most property, the transfer of water rights by lease or sale depends on the clarity of the rights transferred, and for most of the Rio Grande those rights are murky. The costs of purchasing and legally proving the rights are very high, especially in contested transfers. The outcome can be disastrous. In 1999 Intel learned this lesson the hard way when the hi-tech computer chip company discovered that the supposed massive water rights at San Marcial that it tried to purchase didn't in fact exist and $1 million in legal costs gurgled down the drain.[37] A subsequent effort to bypass markets to satisfy water needs in 2009–2010 ended in a similar disaster.

Other problems with water markets also have surfaced. Policing the system of transfers has proved difficult. Lack of oversight has permitted many a farmer to continue using water on his property even after the right has been sold, doubling system depletions instead of improving efficiency of the water use. The public is also slowly realizing that the current policies will eventually lead to the browning of a green river world that is precious to them.

Finally, holders of existing rights do not share a common definition of the economic value of water. Following the lead of the ubiquitous MasterCard credit card advertisements, acequias have decided that their water is "priceless" and have tried to opt out of markets altogether. Meanwhile, desperate municipalities may be able to resort to the power of eminent domain and essentially force some unwilling water rights owners into an involuntary marketplace where courts would settle contested values of water. Everyone has pretty much the same assessment of the value of a pound of copper or a ton of soybeans, but agreeing on the value of an acre-foot of water flowing down the Rio Grande is fundamentally much more difficult.

Water Leasing for the Mimbres

A possible middle option for water markets is represented by the work of a multidisciplinary group that has been exploring the idea of markets for leasing, not selling, water rights.[1] Instead of having to permanently sell the right to divert, a landowner (or water-right owner who does not own land) could decide to lease a specified amount of water (e.g., 10 acre-feet of water per summer for the next three summers to a semiconductor manufacturer) without giving up his legal water right or the ability to use his remaining water allocation. This flexibility could allow for rapid shifts in water use and responsiveness to market forces and to natural water-supply variations without requiring that water be severed from the land. In particular, in time of drought it could allow water to be voluntarily transferred rapidly from a senior user to a very junior one.

At the request of the New Mexico Office of the State Engineer, the research team is developing, modeling, and piloting a water leasing market in the lower Mimbres basin of New Mexico. The Mimbres basin is small (less than one thousand residents), fully adjudicated, and hydrologically closed, with almost all water rights being for agricultural use. As such, implementation of a successful leasing market is much more straightforward there than in a large, unadjudicated and diverse stretch of river, such as the middle Rio Grande.

In the Mimbres market, water trades would be allowed for a limited time (up to five years), thus avoiding permanent shifts of water rights from place to place and use to use. Willing sellers of water rights for the upcoming year would use an online trading system, akin to eBay, to list water rights that they would not need for that year. Willing buyers would bid on them. The system would work like a futures market in advance of the irrigation system and a commodities market thereafter. For example, in March a farmer who intended to plant a high value crop could buy additional rights as a hedge against water shortages later in the season; in August he could also buy additional water for shortages that actually developed.

Some challenges remain in turning the leasing market from a concept to a reality. Many of the same types of legal issues apply as with permanent rights transfers. Other challenges relate to the hydrological and water delivery systems.[2] To apply and police such a water distribution system in a market as large as the Rio Grande would require upgrading the infrastructure and enhancing oversight. Extending the concept of a commodities market to the temporary transfer of water resources may provide many benefits, including avoiding the harsh realities of priority enforcement. But the extension will also require adaptation and flexibility in both economic practice and infrastructure management.

The water leasing concept is a forward-thinking attempt to find "new water" by using what we have more efficiently and flexibly. On the potentially negative side, it represents another step toward the complete commodification of water. In this plan, water would be traded on an open market at the click of a mouse and delivered. As we have seen, this type of valuation of Rio Grande water has, in the past, tended to convert the river into a mere delivery system; in the extreme case the "river" consists of water flowing through pipes or concrete-lined channels to delivery points while the hydrological and ecological river is reduced to a ribbon of desiccated sand. But in this case, the intention is specifically different. The idea is to develop integrated hydrological and economic models that will allow diverse water managers and stakeholders to cooperatively develop viable water markets. The models are intended to avoid negative impacts to the river by identifying potential problems and establishing regulations that help to ensure the market functions in a way that preserves the environment, makes the necessary compact deliveries, preserves traditional practices, and makes money.

Realistically, the leasing market is one of the few proposals that might reallocate water use without destroying the economy of the traditional consumers. If managed successfully, it could help to stretch the flow of the river to accommodate the needs of growing populations without sacrificing what is left of the real river. Can a system intended to establish the value of water as a commodity be successfully structured to also support the value of a river as a river?

Notes
1. Primary investigators are David Brookshire and Craig Broadbent of the University of New Mexico, Vincent Tidwell of Sandia National Laboratories, and Don Coursey of the University of Chicago; see Broadbent et al. (2010).
2. Broadbent et al. 2010.

Store it differently

When one walks into a cellar in Cognac, France, a wonderful aroma compounded of oak and grape and alcohol rises up to meet the visitor. The cognac makers call it the "draught of the angels." Although it is a lovely olfactory experience for the visitor, what the distillery owner smells is the scent of evaporating money. Aging in oak barrels is a necessary part of producing cognac, but as much as one-third of the brandy originally put in the casks seeps through and evaporates. The angels may appreciate the exquisite exhalation from the barrels, but they do not pay money for it.

The Rio Grande, in its own way, also assuages the thirst of the angels. The single largest user of water on the Rio Grande is Elephant Butte Reservoir. When the reservoir is full, the annual evaporation from its surface amounts to about 230,000 acre-feet, and over the past sixty years has annually averaged about 150,000 acre-feet.[38] All of the urban use combined is only about one-quarter of the average evaporative loss.[39] As barrels are required to store cognac, so a reservoir is required to retain runoff in the Rio Grande, but the price is the sacrifice to the atmosphere of almost one-quarter of the water stored. A huge amount of "new" water could be made available if somehow this loss could be reduced.

In retrospect, perhaps Elephant Butte was not the optimum place to build the big storage dam on the Rio Grande. Political tensions between New Mexico and Texas impelled the construction of the reservoir as close to Texas as practical in order to reassure Texan and Mexican users that the water would not be siphoned off during its transit downstream. But Elephant Butte Dam was constructed in the middle of the Chihuahuan Desert. Summertime temperatures exceed one hundred degrees Fahrenheit, humidity is generally very low, and as a result, every year almost eighty inches of water evaporate from the reservoir's surface. Had downstream users been willing to settle for a site as far up the Rio Grande as the present location of Heron Reservoir (7,200 feet elevation) in extreme northern New Mexico, they would have lost only about thirty-six inches of evaporation per year.[40] Rio Grande Reservoir, at 9,500 feet in the headwaters of the river, experiences only about half that loss. Had the big reservoir (or reservoirs) been built high in the mountains, another 100,000 acre-feet per year would have been made available.

Is it too late to rectify the error, to pick new high-elevation dam sites and transfer the water storage duties from Elephant Butte to the new reservoirs? Most likely the answer is yes. All the states and nations involved would have to agree to the new plan and the Rio Grande Compact would have to be amended to allow upstream storage during drought. The multimillion-dollar recreation industry at Elephant Butte would be devastated. Environmental groups would fiercely fight the construction of new dams. What would have been a brilliant idea a hundred years ago is now just a lost opportunity.

In an even more audacious scheme, New Mexico water broker Bill Turner recently took legal action to snatch the angels' draught away from their ethereal lips. In 2003 his company, Lion's Gate Water, filed an application for rights to all of the evaporation from Elephant Butte Reservoir and the other reservoirs in the system.[41] Turner proposed to take the water out of the surface system and store it in aquifers. Cutting off contact with the atmosphere would effectively reduce evaporation to zero. However, many questions arose about this possible solution and it appears to be stalled or dead.[42] The water would (in most cases, at least) have to be pumped into the subsurface, then pumped back out again when needed. Wells are expensive, and a very large injection/withdrawal infrastructure would have to be created to substitute for Elephant Butte Reservoir. All of the pumping requires energy, and energy production requires water. In some ways, the groundwater storage approach is a substitution of energy consumption for evaporative consumption. The legal issues with regard to recovering injected water remain unclear. Groundwater storage does appear to hold promise for reducing evaporative loss, but it seems unlikely that it could provide storage on the scale of Elephant Butte Reservoir.

Value it differently

The cost of water in the West varies sharply and somewhat schizophrenically by use. Water itself is basically free to both farmers and city dwellers, though the costs that are related to it are often hidden or opaque. A considered reevaluation of the true value of this scarce, essential, and public resource could help determine a rational method of valuing it and pricing it.

For urban users in the Rio Grande and in most areas of the West, water is nearly free, with assessed charges covering only the costs of pumping it, treating it, and transporting it. The water itself is assigned no value. Motivated by the spirit of making the desert bloom, water pricing in the West has historically been based on flat rates, independent of consumption, or (even worse) volume discounts have been given to the most profligate users. This situation is gradually changing throughout the Southwest. The region is seeing a shift toward penalties for big users and incentives for water reuse. Water is being priced so as to create incentives to municipal conservation. For example, inverted pyramid pricing structures have been implemented in southwestern cities such as Tucson, Arizona. In this approach, rates increase as water use increases, so that the most conservative users pay the least per gallon of water. Tucson Water also assesses different rates for potable versus reclaimed water, encouraging big users such as golf courses to convert to reclaimed water sources.

What should water cost for a normal city dweller? Consider that the average New Mexican water bill is lower than the average bill for cable and Internet, than the electric and gas bill, or the bill for cell phone and landline. Which is most important? According to Peter Gleick of the Pacific Institute, even water used to fulfill everyday basic human requirements should have a cost. However, Gleick contends that when an individual is incapable of paying for it by virtue of poverty, emergency, or circumstance, this need should be met through community or government subsidies or outright entitlement.[43]

Valuation of water for agricultural uses in the Rio Grande basin is more complex and confounding, demonstrating a wide range of costs and market values. While agriculture still accounts for at least 80 percent of public water consumption in New Mexico, this public water is free to the holders of private water rights who use it for farming. While the water itself is virtually free, the costs associated with its use and the rights to it vary dramatically by geographic location.

On the low end of the cost spectrum, you need go no farther than the Acequia del Molino in Cundiyo, as described in chapter III. In 2009 ditch commissioners there assessed farmers a yearly fee of $2.09 per irrigated acre. That charge covered the cost not of the water diverted from the Rio Frijoles, but the minimal costs of the very simple infrastructure required to get the water to the farms. In a good year, an acre of irrigated Cundiyo land yields eighty bales of alfalfa with a value of $400, although the alfalfa actually is usually consumed by stock on the farm rather than sold. In contrast, in Espanola/Santa Fe water markets, the water rights equal to this one-acre tract in Cundiyo would command upwards of $60,000. Who is to say what the true value of the water is?

A strictly monetary approach to assessing water's value also has its limits. The value of assets not usually equated with dollar amounts has been recognized since the time environmentalists began pointing out the benefits of instream flows to ensure healthy riparian ecosystems. Now economists, biologists, and hydrologists are beginning to more systematically define and assign economic value to both market and nonmarket uses of water, to the ecosystem services that rivers like the Rio Grande provide.

The conceptual framework for the ecosystem services approach to natural resource management has at its heart human well-being as a focus of assessment, rather than attempting to establish an intrinsic value for the natural world, as some environmentalists advocate.[44] Nevertheless, it provides a new dimension to thinking about ecosystems and is a particularly good fit for riparian ecosystems.[45] Based on an economic model, this philosophy blends social and physical sciences to assess and estimate the total value of services provided by an ecosystem. These services include physical and chemical values, such as reducing concentrations of contaminants, but also include perceived values that transcend standard market values, such as recreation, conservation, and other benefits—even ecotourism and *existence value*, the benefit from simply knowing that a certain natural resource exists.

Federal agencies that deal with environmental issues and impacts are embracing this new approach. For example, in summer 2008 the Environmental Protection Agency began an ambitious Ecosystem Services Research program that aims to map ecosystem services throughout the United States by 2014; the U.S Bureau of Land Management is being pushed to move to an ecosystem services approach; and the Agricultural Research Service of the U.S. Department of Agriculture is recognizing the importance of how water affects ecosystem services in arid and semiarid rangelands. These include the production of surface-water runoff, groundwater recharge, and soil water for plant use. These agencies plan to base new soil and vegetation management practices within watersheds on their findings.[46] Even the U.S. Farm Bill passed by Congress in 2008 acknowledges the merits of ecosystem services by facilitating landowner participation in emerging markets for ecosystem services.[47]

Brought to bear on the Rio Grande, this new approach emphasizes what neither municipal nor agricultural policy ever has: the value of water in place. It adds another, and critical, dimension to the already confused debate about what the value of public water is and how that value should shape water policy.

Change management practices

As we have seen, the Rio Grande has suffered in the past from the imposition of short-sighted if well-intentioned solutions to specific problems that in turn created more problems. A recent trend, *adaptive management*, aims for a broader approach, following through management decisions by analyzing impacts and adjusting strategies. Adopted by the U.S. Department of the Interior, this iterative approach requires the participation of managers, scientists, and stakeholders in assessing a problem, designing and implementing solutions, monitoring impacts, and evaluating and adjusting for an improved outcome.

New Mexico took the lead in 2005 in developing the first statewide forest and watershed health plan in the nation, using a landscape-scale approach to achieving and maintaining watershed health and making decisions and taking action at the scale of an entire watershed. This has required collaboration and coordination across political and jurisdictional boundaries that don't necessarily match the watershed, as well as careful assessment of current and historical conditions of the watershed.[48]

Adopting an integrated approach based on good science and the active participation of concerned stakeholders is not easy in an era where patience is in short supply as water managers race to keep up with changing landscapes, reduced water resources, and increasing populations. Yet the results can be remarkably satisfying.

Extreme times call for extreme and flexible approaches to problem resolution. Rio Grande basin water managers can learn from the experience of their counterparts in Australia, who, according to Michael Young of the University of Adelaide, have dropped the term *drought* to describe the extreme climate change and impacts on water resources that they have witnessed in the last several decades.[49] What was considered drought in the past is now acknowledged as the new normal climate. Since 1974 Perth has never again achieved its formerly average streamflow. All formal earlier plans for

Valles Caldera National Preserve

Occupying a hybrid status somewhere between a federal park and a national forest, the Valles Caldera National Preserve (VCNP) in northern New Mexico illustrates the potential for sensible and successful land and water management planning when it is informed by a detailed understanding of a region's climate, geology, soils, vegetation, and hydrology.

Established in 2000, the preserve is a unique experiment in managing public lands through a public-private trust. Essentially, the preserve's mission is broad and ambitious: to provide multiple uses of the land with sustained yield of the renewable resources it contains. These multiple uses include forest management, elk and livestock grazing, watershed management, and the provision of recreational opportunities to satisfy a range of interests, from hikers and ecotourists to hunters and fishermen.

Some consider it free-market environmentalism, others might liken it to ecosystem services on steroids.[1] The objective is to improve forest health, increase streamflow discharge to the Jemez River (a tributary of the Rio Grande), and improve forage for elk and cattle.

In the nineteenth and twentieth centuries, the land of the VCNP region was dominated by extensive logging and sheep and cattle ranching that significantly affected its watersheds and riparian ecosystems. Soils that eroded because of long-term overgrazing by sheep in the 1930s, and by cattle in the 1950s and 1960s, reduced water quality that was further degraded by extensive clear-cutting of timber in the 1960s and 1970s. Second-growth stands of pine, fir, and spruce subsequently sprang up and continue to pose a substantial risk for wildfire and to watershed health. (Changes in the vegetative landscape over a forty-year period are shown in plates 30–32.)

The preserve, formerly the old Baca Ranch, is now a natural laboratory for multidisciplinary research, with the resulting data actively incorporated into forest and range management and modeling. Because this subalpine forest catchment is representative of much of the forest in the southwestern United States, results are expected to be widely transferable.

An example of this research is a project led by Paul Brooks of the University of Arizona-based SAHRA, which established a hydrological observatory in the middle of the VCNP. The project examines how land management decisions, climate variability, and vegetation change affect this unique environment and its associated resources. Through monitoring hydrological, meteorological, and ecological data such as rain, snow, temperature, solar radiation, and streamflow the research has immediate and practical applications. For example, researchers quantified how both thick second-growth forests and completely treeless slopes exacerbate loss of snow and snow water moisture. Nearly 50 percent of the snow water equivalent (SWE) was found to go directly from solid state (ice) to gas (water vapor), and so escape back to the atmosphere before it reaches the ground.[2]

Combined with studies that examined how snow and SWE are distributed around trees, results led to the development of mathematical models that determine how to optimize open space for maximum snow water retention, given a stand's particular age, size, density, and its location with respect to slope, the direction it faces, and elevation.[3] The result is forest thinning prescriptions that are projected to increase stream discharges by 10 to 20 percent.

According to VCNP scientist Bob Parmenter, this means an additional 2,000 acre-feet per year of water that can be used by downstream farmers, ranchers, or urban residents. Translated to dollars, the potential value of the water rights for this saved resource is $2.2 million annually to downstream users at current market prices. The overall value is much greater: additional benefits accumulate from reduced fire risk, improved forest health, and increased wildlife and livestock forage.[4]

Other research keeps track of the complex web of interactions by which the ecosystem feeds back into the water system. The rates at which water vapor and carbon dioxide move from the land surface into the atmosphere, and vice versa, are continuously monitored. The flow of sap up and down through the trunks of pine trees is measured. The movement of nutrients critical to plants and microorganisms is traced throughout the landscape.

The researchers take advantage of fluctuations in the natural system to understand how water, energy, and nutrient use respond to changing conditions. For example, a delay in the timing of snow accumulation in the fall allows soils to freeze, changing the way that water is distributed and dramatically slowing forest growth the following spring.

After several years of monitoring, spruce budworms infested and devastated the pine trees at the study site. The researchers considered this an opportunity rather than a disaster. Large tree die-offs are expected in the coming decades as the climate warms and dries.[5] The tree death at the VCNP will allow researchers to study how other vegetation responds to this loss of trees, whether water use changes, how snow accumulation is affected, and how the nutrients are rerouted. Although the prospect of dying forests is not appealing, the data gathered will improve understanding of such fundamental issues as how water supplies might change, whether erosion will increase or decrease, and how the buildup of carbon dioxide in Earth's atmosphere might be affected. The study will run for many years, supplying data and stimulating hypotheses that will help to guide society's response to a rapidly changing environment in the southwestern United States.

Notes
1. Yablonski 2004.
2. Musselman, Molotch, and Brooks 2008.
3. Veatch 2008; Rinehart, Vivoni, and Brooks 2008.
4. Parmenter 2009, 24.
5. Van Mantgem et al. 2009.

managing the Murray-Darling Basin in southeastern Australia have been suspended, because they were simply incapable of addressing the extreme new realities of a radically changed climate.

According to Young, governance needs to be based on expertise and aimed at facilitating change as fast as system shifts occur. In Australia, every state has written new legislation that redefines property rights and, therefore, water rights. Basically, these reforms consisted of abandoning the riparian form of water law that had prevailed ever since the beginning of English settlement in the early nineteenth century, in which water ownership is associated with land ownership adjacent to water bodies. Instead, public ownership of water was asserted, leading to a system of transferable water licenses for specific amounts of use. In consequence, a water-rights market has developed.[50] After one hundred years, Australian water law is finally catching up to Morris Bien. With the potential of catastrophic water resource shortages throughout the western United States, drastic reevaluations of water and land management practices are also in order that focus not only on mitigation, but also on adaptation, with everything on the table including removing barriers to trade, and expediting adjustments in the water supply, water allocation, and water transport.

Limit growth based on the water supply

"Our goal is to not have water be a constraint to growth. We don't want to be ones to say 'Yea' or 'Nay' to growth."[51] So said Steven Robbins, chief engineer for the Coachella Valley Water District in Southern California. He was responding to a 2007 report that aquifer depletion and consequent land subsidence will force a reduction in pumping from the aquifer that supplies water for Palm Springs and the rest of the Coachella Valley. In spite of extreme desert conditions (less than three inches of rain each year), the population of the area grew from 300,000 to 400,000 between 2000 and 2007 and is projected to grow to 1 million by 2060.[52] All of this growth was, and is projected to be, sustained by pumping of nonrenewable groundwater.

Robbins's statement highlights the current head-to-head collision of values and visions for the future of the American West. Is sustainability a reasonable societal goal? If so, then what constraint to growth could be more appropriate than the amount of water available to sustain human population over the long haul? Or is growth for its own sake a goal so compelling that all other considerations must give way before it?

The economic argument for growth is that it fuels the economy to grow faster than the population and thus provides benefits to both old and new residents. The same argument cannot be applied to the water supply. The water supply of the Rio Grande valley is fixed and it is fully used. The redistribution of the water supply is a zero-sum game. If large numbers of new residents move to the valley, the water they consume must be taken from either the ecosystem or the residents who preceded the newcomers. Redistribution can either be accomplished by reducing per capita urban use or by putting farmers off the land.

What are the benefits of this redistribution to the current residents of the valley? Certain sectors—land developers and the construction industry, for example—benefit directly and often very significantly from population increases. Whether benefits to other current residents outweigh the costs in terms of lost quality of life is a matter of debate.[53]

One approach to achieving sustainability for the Rio Grande would be to turn the usual societal equation on its head. The question we should be asking is not *How can we continue to have a living river while meeting the water supply demands necessary for growth?* but *What uses do we want to see the waters of the Rio Grande put to, and how much to each?* The answer might be that current residents value the agricultural ambiance and natural environment supported by the water more than they do the economic impacts of using the water to support continued growth. Were this so, one course of action might be to pass legislative restrictions on water rights transfers or on the allocation of water to various economic sectors. Such policies could have the effect of greatly restricting growth.

This would represent a radical departure from both the long legal history of water rights in the region and from current attitudes toward growth. Given the enormous financial and political influence of the economic sectors that benefit from growth, enactment of such restrictions in the foreseeable future is highly unlikely. Nevertheless, even opening a broad-based public conversation on the topic might prove beneficial.

Chapter XI
The Future of an Old River

Is the river a river, or is it a conduit for a liquid commodity? If it is the latter, the logical end product will be a Rio Grande flowing only in a network of steel pipes beneath a valley carpeted by low-density housing tracts. The future of the river depends largely on the attitude of those who live along its banks and on the political structures that act on that attitude. For the river to reign in the Rio Grande valley, priorities and values must be realigned.

The early humans and the Pueblo tribes who inhabited the river valley were very patient people by our standards, yet their lives were just a flicker compared to the slow evolution of the ancient river. Today, we want the river to do our bidding. Vast populations now depend on it. We demand water deliveries under the Rio Grande Compact, allowing no more than a few years of shortfall. No matter how torrential the rain, the river must stay within its assigned limits. Farmers have to receive their next irrigation allotment before their crop starts to wilt. There is an inherent conflict between the timescales and the processes that created the river and the pace and needs of the humans who now inhabit its valley. People have consequently altered the entire complex: the hydrological river, the ecological network that surrounds and depends on it, and the landscape around it.

The Human Mirror

The history of the human occupation of the Rio Grande watershed proves this fundamental fact: water reflects. Changes to the river show as much about people living near it as they do about the river itself.

The dangers of living in a stressed environment are brought home by two previous civilizations in the same region: the Chaco Anasazi and the Hohokam. Both slowly developed societies far more sophisticated in terms of massive engineering projects and alteration of their surroundings than the groups from which they arose. In spite of apparent long-term stability, neither of these societies was sufficiently resilient to withstand the environmental stresses from a few years of unusual, but not unexpected, fluctuations in weather. Both left behind complexes of ruins that still amaze today's tourist.

Short of a global collapse of civilization, today's society on the Rio Grande probably will not disappear virtually overnight as did its ancient predecessors, but it certainly is vulnerable to disruption and decline. The Achilles' heel of the southwestern United States is the scarcity of water. Stretching the existing supply through conservation and switching from irrigation to industrial and municipal uses as populations increase can only go so far. Although, in theory, large amounts of water can be made available by shifting consumption from agriculture to other users, constraints ranging from the Rio Grande Compact to the Endangered Species Act actually leave very limited room to maneuver.

In fact, the current and future residents of New Mexico and the middle Rio Grande, in particular, will have to pay for many prior years of groundwater pumping whose effects will be felt for years to come. Groundwater pumpers have to keep buying up water rights to retire to offset the delayed effects of past pumping so that compact requirements can be met. What rights are left to purchase? Only those of farmers. But the balance sheet is untenable. To offset the pumping rights that already have been permitted, water rights to around 57,000 acres of farmland in the middle Rio Grande would have to be surrendered, according to State Engineer John D'Antonio. Compare this with the 50,000 to 65,000 acres estimated to comprise the MRGCD reclaimed lands, the Pueblo lands, and the land that was irrigated before 1907.[1] Something has to give. The question is not whether the Rio Grande valley of one hundred years from now will be habitable—it almost certainly will be—but whether it will be a place where people will want to live?

Yet more dues will be paid to offset the environmental impacts that date back to the arrival of the industrial revolution in New Mexico. The entry of the railroad, the opening of new markets, and the ability to import heavy machinery enabled a surge in natural resources exploitation. Compared to the conservative pace of agriculture, ranching, and logging as practiced by the Pueblo and Hispanic farmers, the late nineteenth-century newcomers saw themselves in a competitive struggle to strip the land of its resources.

Everything about the newcomers was highly mobile and very temporary. They brought capital and technology to a world that had never known it. In contrast to the earlier inhabitants, they had no long-term stake in the land they claimed. Their goal was simply to temporarily gain control of the land so as to squeeze the resources from it before their competitors did. Once the land was depleted, they generally moved on.

The health of the modern Rio Grande and its landscape still reflects this same tension between conservative values and the desire for change. Although the ability to generate wealth no longer constitutes a self-evident justification to the extent that it did during the Gilded Age, the ideal of "growth" still generally needs no explanation or apology. The future of the Rio Grande valley will depend to a great extent on whether the prevailing voices are those of individuals who will ultimately profit handsomely from great projects that contribute to growth or are those of residents who are concerned about the long-term sustainability of the valley.

Counterweights to Rapid Change

New Mexico's legal codes and compacts play a paradoxical role in the life of the Rio Grande. The doctrine of prior appropriation originally aimed to promote the most rapid possible exploitation of the river. The person who first managed to secure control of the water got the assurance of control in perpetuity. In accepting Nathan Boyd's application for a right-of-way to build a dam across the Rio Grande at Elephant Butte, the federal government was poised to carry this philosophy to its logical limit. Because he was able to raise the capital necessary to harness the river and was sufficiently aggressive to get there first, Boyd was nearly granted control of the entire river. Subsequently, both the federal and state governments stepped back from taking prior appropriation to its logical conclusion and recognized that the river was a shared resource of the entire region. The Rio Grande Compact later formalized the principle that all consumers of the river should get an equitable share of its water, rather than giving control of its distribution to the first individual able to physically control its flow.

The paradox arises from the insulation from economic forces that New Mexico's water code provides for existing users. Although as a public resource water must be put to beneficial use, the code contains no ranking of how beneficial the various uses must be. Uses with relatively low economic return (e.g., alfalfa farming) have just as much right to the limited and public water as do very remunerative uses (e.g., semiconductor manufacturing). Fees paid for the water are based on the cost of diversion, not on the potential economic value of the water. The water code has thus tended in the long term to actually slow economic development in the region while also helping somewhat to preserve the traditional riparian landscape and ecosystem of the valley. Similarly, the Rio Grande Compact has ensured that the river flow cannot be siphoned off upstream by some claimed "highest use" and thus has ensured a living river all the way south to El Paso.

Steve Reynolds's integrated regulation of river and groundwater consumption in the middle Rio Grande was designed to ensure that New Mexico met the demands of the compact, but the unintended consequence has been to also maintain a wet river. Environmentalists may sometimes bemoan the protection afforded to the "inefficient" agricultural sector's large consumption of Rio Grande water, but the net effect of these nearly one-hundred-year-old legal constraints has been to help maintain a vital river habitat. In any case, the interests vested by this legal framework are so large and entrenched that significant change is unlikely. Efforts at reform may be better spent in adapting the existing framework rather than attempting to replace it.

Is the Solution More Technology?

Beginning in the mid-nineteenth century, humans began to substantially alter the Rio Grande and its watershed. Massive diversion works were constructed in Colorado and innumerable dikes and dams popped up in New Mexico. Huge numbers of livestock were introduced. Forests were stripped to feed burgeoning industry. Inevitable consequences followed: water supplies become more erratic, the riverbed rose, fields were salinized, levees failed during floods. Two major initiatives were launched to deal with these problems: (1) the formation of the irrigation districts; and (2) the reengineering of the river. Both of these initiatives relied on technological fixes. The irrigation districts focused on construction of massive levees, elaborate diversion works, and a network of drains. The U.S. Army Corps of Engineers attempted to convert the river into an efficient transmission line for water, straightening and channelizing it, even excavating it far below the water table in order to suck back the water lost by seepage from the aggraded original channel.

With each technological fix, the original Rio Grande receded a step and so did its intrinsic ecological and aesthetic values. By the 1960s the straight, narrow Rio Grande, confined between high banks and surrounded by towering levees and dense thickets of saltcedar, was seen by the citizens of Albuquerque only as a fit site to dump their refuse. Today, the general consensus is that reliance on technology went much too far. In line with river restoration efforts throughout the country, more jetty jacks are being removed with cranes, and levees are being relocated in order to expand wetlands.[2] Modern technological tools such as computer models of river flow, remotely controlled headgates, and instantaneous satellite images of river conditions have aided greatly in managing a better-functioning river. Projects that substitute massive engineering works for river management, however, are likely a thing of the past.

Perhaps what is needed more than many small projects to expand wetlands or increase high flows is a fundamental change in attitude toward the river by the residents of its valley. The technological fixes were the product of a mindset that asserted that the river must be forced to conform whenever human activity and the behavior of the river came into conflict. If the river flooded, the levees had to be raised. If it failed to deliver the required water, it had to be channelized. We can today take a lesson from those earlier inhabitants of the valley who built their villages on high ground to avoid flooding and otherwise accommodated their activities to the natural course of the river, as well as the fate of those who were heedless of it. The laws of hydraulics cannot forever be held at bay. The longer the river is confined behind levees, the higher it will aggrade, until it becomes a raised causeway high above the floodplain. When the levees fail, as they inevitably must, the disaster will be all the worse. Perhaps someday society will acknowledge this by purchasing land for a new river path and intentionally creating a new, low route for the river before it accomplishes the same by force. Massive technological fixes are the product of a one-way mindset. Enjoying a living river requires an attitude of reciprocity.

The Rio Grande Tomorrow

Many changes for the better have come about since the Rio Grande's nadir in the late 1960s. The river is now valued by many for its own sake, not just as a channel for transporting an economic commodity. As a result, it is beginning to look and feel at least a little like the river it once was. Nevertheless, there is no assured bright future. Many forces are working in the opposite direction.

Perhaps the most worrisome change is the one furthest out of the control of the residents of the Rio Grande: the global climate. The apparently inevitable and (on human timescales) irreversible drying and warming in the Southwest will negate much of the current success in water conservation.[3] These changes will heighten competition and suspicion among competing water users in the drainage basin. Water to maintain a "wet" river will become far more precious.

Even if the climate were stable, managing the system within the confines of the present legal strictures would become increasingly difficult. Water-balance modeling studies show that New Mexico will be hard-pressed to consistently meet the requirements of the Rio Grande Compact.[4] Albuquerque has already affected the flow of the Rio Grande by switching from groundwater pumping to San Juan–Chama water in the river. In one respect, this is a positive change, since surface water is a renewable resource. However, in the past, San Juan–Chama water provided a safety valve in the form of a flexible pool of water that could be reallocated to emergency uses such as silvery minnow flows. Now that pool will no longer be available. In the longer term, a growing population will outstrip the San Juan–Chama flows, and heavy pumping will have to resume. Additional surface rights will have to be purchased, resulting in loss of farmland and a dramatic change in the character of the communities that line the riverbanks. As the impacts of global warming decrease the flow of the river, its bed will continue to aggrade. As the years go by, it will become increasingly difficult to keep the river wet. Waterlogging in the adjoining lands will increase, and the river will pose an increasing threat of catastrophic flooding. Unless the inhabitants of the valley are willing to plan far ahead and invest the necessary resources now, the present upswing in the health of the river may prove to be illusory.

Restoring Balance to the Rio Grande

This book started with the ancient people of Cochiti Pueblo, a people with a long-lasting respectful relationship with the river who now reside next to the massive, modern structure that controls it. Many residents of the Rio Grande find themselves in the same situation. A river that was once capricious, powerful, even sacred, has been increasingly converted into an object for the fulfillment of human needs. Viewed alternately as a necessity and a threat, the river has been surrounded and controlled by engineering works and restricted by a network of laws and compacts until the physical reality— flowing water, fish, and trees—recedes into insignificance behind the claims, rights, and decrees.

It has been said that "where there is no vision the people perish."[5] The life of the Rio Grande depends on the long-term vision of the people around it and their willingness to implement that vision. A sustainable and at least semi-natural river will depend on a vision that goes beyond specifics such as an engineered refugium for silvery minnows or modifications to the flow regime. It is the big decisions, or perhaps more pessimistically, the lack of big decisions that will in the long term determine the fate of the river. Al Utton, the late University of New Mexico professor who devoted most of his public and private life to teaching about water and transboundary resource issues, liked to say, "Education by catastrophe is the way that we run western water resources. We must stop proceeding that way."[6] What are the priorities for the use of water and who gets to decide? How will current residents benefit from adding more population? How long can the valley continue to grow without changing into something current residents would rather not see? Are both an indefinitely sustainable economy and a natural Rio Grande mutually achievable?

For most of the twentieth century, the inevitable tension between short-term economics and long-term natural processes, between the interests of those already here and those who wish to come, between stability and growth, seesawed back and forth but predominantly was resolved to the detriment of the river. And yet, there has been progress. The Rio Grande still flows on through New Mexico, sustaining increased populations, new industries, and local farming. Boyd's bid to gain personal control of most of the Rio Grande–irrigated farmlands of southern New Mexico and western Texas was thwarted. Agriculture held back the complete commodification of water. For all its false starts, the environmental movement has persisted in promoting a broader perspective of watersheds and riparian habitats, and in pointing out the value of cooperative river management for concerns that transcend the solely human. A sustainable Rio Grande now requires human intervention and action, the continued balancing of interests, and a commitment to facing rather than ignoring challenges to the mighty river.

To the ancient peoples who lived, and some of whom still live, on its banks, the Rio Grande was a mysterious and especially powerful part of a world that was inherently sacred. For many who arrived much later, the river itself meant little. It was valued only insofar as its water represented an opportunity for the individual to achieve economic advancement. The flip side of the opportunity was the threat that the untamed river posed to property and life. In response to opportunity and threat, the river was reined in. In places it was reduced to a ruler-straight current between the steep walls of a ditch.

Standing back and surveying five hundred years of human interaction with the Rio Grande, we cannot adopt either of these endpoints as a basis for moving into the future. Too many people depend on using the waters of the river for it to ever be unharnessed. Except for high in its headwaters, it will never again be a wild river. And yet, when they stand on its banks, most of the inhabitants of the Rio Grande valley see much more than just an economic resource. In common with the earliest inhabitants, they see a living environment that has inherent beauty and worth. The challenge of reining in the Rio Grande is over. The challenge of tomorrow is achieving a peace with a river that both nature and humans can abide.

Notes

Prologue

1. Pecos 2007.
2. Lange 1959, 77–78.
3. Scurlock 1998, 18, 22–23.
4. Lange 1959, 78–79; Scurlock 1998, 94.
5. Spencer 1947.
6. Dominguez 1956, 159.
7. Bureau of Indian Affairs, Albuquerque Area Office, Branch of Real Estate Services, *Pueblo Land Status*, February 1, 1979 (unpublished), 31–32; Bowden 1969, 1224.
8. These included land litigation between the Cochiti Pueblo and Hispanic intruders (*Cochiti Pueblo v. C de Baca et al.*, 1817), as well as Cochiti Pueblo struggles with neighboring Pueblos (*Cochiti Pueblo v. Santo Domingo Pueblo*, suit over land, 1817); see Twitchell (1914/2008, vol. 1; vol. 2, reel 6, frames 30–75).
9. Bureau of Indian Affairs, *Pueblo Land Status*; Hall 1984, 1987.
10. Hall 1984, 114–20; *United States v. Lucero*, 1 N.M. 422 (1869).
11. Hall 1984, 197–207.
12. Article XVI of the 1938 Rio Grande Compact; Act of March 13, 1928, the MRGCD Act authorizing participation on terms by the six middle Rio Grande pueblos (Public Law 69, 70th Cong. [S 700; March 13, 1928]). See Burkholder (1928, 102–4).
13. Burkholder 1928, 121–22; table 5, p. 44.
14. Vlasich 1980, 25.
15. 1948 Flood Control Act; Specific construction of Cochiti authorized by the Flood Control Act of 1960 (Public Law 86-645, 74 Stat. 488); see Kelly et al. (2007).
16. This account draws heavily on Pecos (2007).
17. *Pueblo de Cochiti v. United States*, 647 F. Supp. 538 (1986), filed November 29, 1985.
18. Bensfield and Hall 1970.
19. Remarks by Pecos at Utton Center Conference on New Mexico Dams and Storage on the Rio Grande, University of New Mexico Law School, Albuquerque, October 15, 2007.
20. Kelly et al. 2007.

Chapter I

1. Karlstrom et al. 2004.
2. Mack 2004a; Blakey and Ranney 2008, 14–20.

3. Blakey and Ranney 2008, 103–19; Cather 2004.
4. Chapin, McIntosh, and Chamberlin 2004.
5. Bindeman 2006.
6. Miller and Wark 2008.
7. G. Smith 2004; Connell 2004; Mack 2004b.
8. G. Smith 2004; Connell 2004; Pazzaglia and Hawley 2004; Hawley 2005.
9. G. Smith 2004.
10. Siebenthal 1910, 112–14.
11. Tuan 1973.
12. Ibid.
13. Kurc and Small 2004.

Chapter II

1. Poore 1894, 438.
2. Barrett 2002, 2, 115–16.
3. Martin 2005; Kerr 2010.
4. See Haynes Jr. 1991, 447; 2008; and Kennett et al. 2009.
5. Stuart 2000, 16–17.
6. Some archaeologists believe instead that there was cultural continuity between late Paleoindians and Early Archaic peoples in the Rio Grande area and elsewhere in New Mexico.
7. Mabry 2005, 53.
8. Ibid., 54.
9. Riley 1995, 59.
10. Mabry (2005, 55) cites a number of sources supporting this, including Cummings and Moutoux (2000); Fish (1998); Huckell (1995, 1998); and B. Phillips (2000).
11. Plog and Gray 1997, 68; citing Kaplan 1965.
12. Wills 1992, 160.
13. Anschuetz 1998, 132–33.
14. Adopting the more useful four-part classification by Anscheutz (1998, 139–52) of Pueblo agricultural systems, which also apply to the other major cultural groups of the Southwest.
15. Lawton and Wilke 1979, 33; Vivian 1974.
16. Wozniak 1997, 6.
17. Mabry 2005, 60; Nials 2008, 168.
18. Damp, Hall, and Smith 2002, 665.
19. Graybill et al. 2002; Abbott 2003.
20. As discussed in Anschuetz (1998, 132–33) and Cordell (1984, 203–4). The problems associated with stagnant water were not unique to the Rio Grande, however; Cedric Kuwaninvaya of Sipaulovi Village of Hopi points out that Homol'ovi IV near present-day Winslow, Arizona, was abandoned because of a mosquito infestation in the nearby Little Colorado and resultant malaria outbreak. After subsequent clan migrations throughout the area, the village of Sipaulovi was established on Second Mesa. Acknowledging the clans' tribulations, the village name means "place of the mosquito."
21. Vlasich 2005, 7.
22. Stuart 2000, 57.

23. Studies at Newcomb, New Mexico, show an extensive agricultural system may have been built in the Two Grey Hills area of the San Juan basin by pre-Columbian Chaco Anasazi and involved the use of irrigation ditches and efficient diversion of tailwater to a series of fields as long ago as AD 500; see Friedman, Stein, and Blackhorse Jr. (2003).
24. Lawton and Wilke 1979, 33.
25. Stuart 2000, 66.
26. Lekson 2006, 11.
27. Vivian 2004, 11.
28. Force et al. 2000.
29. Sebastian 2006, 396.
30. Lekson 2006, 15.
31. Other theories include extreme vegetation change from overlogging, leading to a scarcity of fuel for cooking and warmth; see Betancourt, Dean, and Hull (1986) as referenced in Diamond (2005, 147); an elaborate and complex social system that depended on outlying communities for crops and the redistribution of goods, ending in starvation for many when the system collapsed (Diamond 2005, 143) and cutoff of support for and unequal distribution of resources and goods to the farmers (Stuart 2000, 124).
32. I. Clark 1987, 6.
33. Sebastian 2006, 401.
34. Diamond 2005, 156.
35. Stuart 2000, 115; citing also Cordell 1997, 318–19; McKenna 1984; and Akins 1986.
36. Stuart 2000, 113; citing Akins 1986, 61.
37. Solstice Project 1999.
38. Cordell 1994, 130–33.
39. Stuart 2000, 139–40.
40. Benson and Berry 2009.
41. Cordell 1994, 133.
42. For an extended discussion of this movement and population aggregation, see Adler, Van Pool, and Leonar (1996, 418).
43. Stuart 2000, 153.
44. Some archaeologists maintain that numerous Puebloan ancestors lived in the middle and upper Rio Grande before the Mesa Verde area was abandoned; see, for example, Lakatos (2007) and Boyer et al. (forthcoming). Demographic changes in the northern Rio Grande area, however, were dramatic.
45. Wozniak (1997, 6–7) contradicts F. Ellis (1970, 1979) on the basis of insufficient evidence. F. Ellis and Dodge (1989, 51) purport that descendants of Mesa Verde or Chaco spread the idea and technology of irrigation down the Puerco and eastward to the Rio Grande. There is some debate also among the archaeological community as to whether Taos Pueblo was irrigating prior to the arrival of the Spanish.
46. Stuart 2000, 150; citing Stuart and Gauthier 1986.
47. At farmlands located along small permanent streams or marshes, near Bernalillo, at Zuni and Acoma, for example. Wozniak (1997, 10–11) provides a summary of the Spanish expeditions that noted irrigation practices, including Rodriguez-Chamuscado, Espejo, and Sosa.
48. Wozniak 1997, 12.
49. Stuart 2000, 162; as per Dozier 1970, 24; Whitely 1982; and Cordell 1994, 15.

50. Emory Sekaquaptewa, taped interview by Mary E. Black, November 2007.

Chapter III

1. Villagrá Alcalá 1610/1933, 126.
2. Anon. 1598/1953.
3. Bullard and Wells 1992.
4. Van Cleave 1935.
5. Sublette, Hatch, and Sublette 1990.
6. Westphall, 1983, 3-5.
7. Baxter 1997, 1-2.
8. Glick 1970, 14; Meyer, 1984, 19-23.
9. Westphall 1983, 19; Hall 1984, 4-5; Ebright 1994, 23.
10. Hall 1993, 94-95.
11. Ibid., 93-94.
12. For example, see Baxter (1990, 42-70).
13. Hall 1993, 96; *Upper Rio Grande Hydrographic Survey, Santa Cruz River Section, 1963 (1990 rev.)*, "Map Sheet 16: Hydrographic Survey of Cundiyo, Rio Frijoles" (Santa Fe, NM: Office of the State Engineer).
14. Jenkins and Baxter, 1986, 18-25.
15. *Upper Rio Grande Hydrographic Survey,* "Map Sheet 16."
16. Sabino Samuel Vigil, interview by G. Emlen Hall, Cundiyo, NM, March 21, 2009.
17. *Brief of Amicus Curiae New Mexico Acequia Commission* (March 13, 2008) in *Pena Blanca Partnership et al. v. San Jose de Hernandez Community Ditch et al.*, 202 P.3d 814 (2008).
18. Chama River: *State of New Mexico ex rel. State Engineer v. Aragon et al.*, U.S.D.C. N.M. Civ. No. 7941 S.C.; re Santa Cruz: *State of New Mexico ex rel State Engineer v. Abbot et al.*, U.S.D.C. N.M. Civ. No. 7488 and 8650 consolidated S.C.
19. Simmons 2003, 13-15.
20. *Application of the Middle Rio Grande Conservancy District to Change Points of Diversion*, No. 0620, filed March 18, 1930, Office of the State Engineer, Santa Fe, NM; Follett 1896, tables 12-13.
21. M. Simmons 1968, 1969, 1972.
22. Rodriguez 2006, 75-81; S. Crawford 1988, 31-78.
23. S. Crawford 1988.
24. Ibid., 87-90.
25. See Glick (1970).
26. Rodriguez 2006, 75-101; Steele 1994, 141-69.
27. Taylor 1975; Orona 1999, 85, 39n.
28. Orona 1999, 146.
29. Baxter 1997, 30, 48; Ebright, 1994; Orona 1999, 86.
30. Baxter (1990, 23-70) details the interacequia disputes in the Taos Valley; Follett 1896, 74; Orona 1999, 64.
31. Hall 1983, 15-20.

Chapter IV

1. See section 1, "Water Courses, Stock Marks, etc.," Kearny Code of Laws, adopted in

Santa Fe, New Mexico, September 22, 1846, and compiled in pamphlet 3, "Territorial Laws and Treaties, Historical Documents," vol. 1 of *New Mexico Statutes Annotated 1978.*
2. Baxter 1997, 86.
3. O'Sullivan 1839, 426–30.
4. Wilber 1881, 143
5. Hayden 1869, 141.
6. Gregg 2001, 349.
7. Isaiah 35:1–2.
8. U.S. Bureau of the Census 2009.
9. Newell 1893, pt. 3a, 7–99.
10. Myrick 1990, 4.
11. Marshall 1945, 138.
12. Ibid., 139.
13. Myrick 1990, 109–17.
14. Ibid., 60.
15. Scurlock, 1998, 128.
16. R. Reynolds and Pierson 1942, 9–10, 17–18.
17. DuBuys 1985, 9.
18. Ebright 1994, 3–11, 263–72.
19. Ibid., 145–68.
20. *Treaty of Guadalupe Hidalgo: Findings and Possible Options Regarding Longstanding Community Land Grant Claims in New Mexico* (Washington, D.C.: GAO), table 14, p. 93.
21. DuBuys 1985, 228.
22. Allen et al., 2002.
23. Dortignac 1956, 57.
24. Ibid., 57.
25. MacCameron 1994.
26. Allen et al. 2002.
27. Cooperrider and Hendricks 1937.

Chapter V

1. Follett 1896, 163, 170–71.
2. Burkholder 1928, 26.
3. Carlson 1967, 115.
4. V. Simmons 1979.
5. Machette, Marchetti, and Thompson 2007.
6. Thomle 1948, 78–79, 88–91.
7. Follett 1896, 24–26, 46–49, 159–60.
8. Ibid., 24.
9. Thomle 1948, 88.
10. Values are from Follett (1896, 47–48).
11. Thomle 1948, 78.
12. Follett 1896, 159–60.
13. Ibid., 168–69.
14. Ibid., 135.

15. Ritter, Kochel, and Miller 2002, 195–99.
16. Ibid., 204–5.
17. Ibid., 224–25, 235.
18. Scurlock 1998, 33–37.
19. Compiled by Scurlock (1998, 33–37).
20. Follett 1896, 163–65; Cooperrider and Hendricks 1937.
21. Sargent and Davis 1986, 100.
22. Wozniak 1997, 99; Harroun 1899, 25.
23. S. Mills 2004, 106.
24. Follett 1896, 127–31; Burkholder 1928, 34–55; Natural Resources Committee 1938, 70; Hedke 1925.

Chapter VI

1. Baxter 2000.
2. Frazier and Heckler 1972.
3. Ibid., 8.
4. Malone 1934, 456–57.
5. Frazier and Heckler 1972.
6. From Murphy (1904); a photograph of the actual current meter built for the 1889 campaign can be seen in Frazier and Heckler (1972, 15).
7. Follett 1896.
8. Ibid., 184.
9. Baxter 1997.
10. Ibid.
11. See *Snow v. Abalos*, 18 N.M. 681 (1914); and I. Clark (1987, 103–4).
12. *Candelaria v. Vallejos*, 13 N.M. 146 (1905).
13. See *Acequia de Llano v. Acequia de las Joyas del Llano Frio*, 25 N.M. 134 (1919).
14. Powell 1879.
15. I. Clark 1987, 60; Pisani 1992, 150–51.
16. Rowley 2006.
17. Pisani 1992, 301–2, 306.
18. Littlefield 1987.
19. Littlefield 1987, 44–49.
20. I. Clark 1987, 92.
21. 55th Cong., 2nd sess., S. Doc. 229, 14–18; Boyd 1902.
22. Lester 1977, 51–53.
23. U.S. Reclamation Service 1903, 377–78; 1904a, 395–415.
24. From the 57th Cong., 2nd sess., S. Doc. 154, 24, quoted in I. Clark (1987, 93).
25. I. Clark 1987, 95.
26. Quoted in I. Clark (1987, 715n53).
27. *United States v. Rio Grande Dam and Irrigation Company*, 174 U.S. 690 (1899); 184 U.S. 416 (1902); 215 U.S. 266 (1909).
28. I. Clark 1987, 96.
29. U.S. Reclamation Service 1903, 33.
30. Pisani 2002.
31. Bates 1957.
32. Frazier and Heckler 1972, 23.

33. U.S. Reclamation Service 1904b, 317. See Hall (2008a). For use of the term *enginoir,* see Hall (2002a, 132–33).
34. Bien 1905, 29–34.
35. Merryman 1985.
36. See section 73, An Act to Conserve and Regulate the Use and Distribution of the Waters of New Mexico, chapter 49, 37th Legislative Assembly, *1907 Session Laws of the Territory of New Mexico,* 71–95.
37. From Hall (2008a, 247); An act entitled an Act to Provide for a State Irrigation Code, chapter 132, *Laws Passed at the Ninth Session of the Legislature of the State of South Dakota,* March 3, 1905; Irrigation Code, chapter 34, *Laws Passed at the Ninth Session of the Legislative Assembly of the State of North Dakota,* March 1, 1905.
38. See section 22–25, An Act Creating the Office of the Territorial Irrigation Engineer, chapter 102, *1905 Acts of the Legislative Assembly of the Territory of New Mexico, Thirty-Sixth Session,* March 16, 1905.
39. An Act to Conserve and Regulate, chapter 49, 37th Legislative Assembly, *1907 Session Laws of the Territory of New Mexico,* 71–95.
40. The U.S. Bureau of Reclamation has rights to water stored behind Elephant Butte Dam under two state permits, one secured from the territorial engineer in 1907, and the other secured from the new state engineer in 1915. The 1915 permit secured for the United States all the then-unappropriated water in the Rio Grande under state law.
41. From Court of Appeals Brief.
42. *Community Ditches or Acequias of Tularosa Townsite v. Tularosa Community Ditch,* 16 N.M. 750 (1911); Hall 2000.

Chapter VII

1. Autobee 1994.
2. Granjon 1986, 44; as cited in L. Harris 1996, 9.
3. Schönfeld LaMar 1984, 38, 102, 108; citing U.S. Reclamation Service 1917, 61; and Bloodgood 1921, 10–17.
4. Schönfeld LaMar 1984, 103; Walker and Curry 1928, 81.
5. Rio Grande Project 1917, 82–86; U.S. Bureau of Reclamation 1972, 393; as reported in Autobee 1994.
6. Schönfeld La Mar 1984, 160.
7. L. Harris 1996, 10; as per Lester 1977, 92.
8. Autobee 1994.
9. Calkins 1936; see also Schönfeld LaMar 1984, 158.
10. Schönfeld La Mar 1984, 212.
11. Ibid., 258.
12. Leopold and Koelzer 1947, 40; U.S. Bureau of Reclamation 1983, 1, 6; Simpich 1920, 65.
13. Calkins 1936.
14. *Bernalillo County Survey,* "Map Sheet 33" (Albuquerque, NM: District 1, Office of the State Engineer, 1916), by County Surveyor Pitt Ross, November 30, 1916, retraced December 1933 and February 1934 by C. B. Breyer, County Surveyor.
15. *Grolet-Gurulé: Los Franceses de Nuevo Mexico 2009,* http://www.GuruleFamily.org (accessed September 7, 2009).
16. Sargeant and Davis 1986, 14–16; see also Simmons 2003, 90–91.
17. Sargeant and Davis 1986.

18. "Map Sheet 33, Drainage Investigations" (Albuquerque, NM: District 1, Office of the State Engineer, 1917), pursuant to chapter 71, *Laws of New Mexico*, approved March 13, 1917.

19. Gault 1923; Wozniak 1997, 130.

20. Fox 1928.

21. Davis 1857, 195.

22. Hays 1959; Orona 1999, 58–60.

23. Orona 1999, 152.

24. Ibid., 131–40, 161.

25. Ibid., 128.

26. *In re the Proposed Middle Rio Grande Conservancy District*, 31 N.M. 188, 190 (1925).

27. Orona 1999, 179–86.

28. *1927 Laws of New Mexico*, chapter 45, 135–207; I. Clark 1987, 210–11.

29. Middle Rio Grande Conservancy District, "Statement to Accompany Application 0620 by Middle Rio Grande Conservancy to State Engineer for Permits to Change Points of Diversion and Place of Use of Certain Water Rights in and to the Water Rights of the Rio Grande in the Counties of Sandoval, Bernalillo, Valencia, and Socorro in the State of New Mexico, November 25, 1930" (Albuquerque, NM: Archives, MRGCD).

30. "Status of Community Ditch Agreements with the Middle Rio Grande Conservancy District as of September 1st, 1939" (listing agreements with fifty-two acequias, primarily dated in the early to mid-1930s; Albuquerque, NM: Archives, MRGCD).

31. For example, the agreement between the MRGCD and Los Padillas Acequia, June 10, 1935 (Albuquerque, NM: Archives, MRGCD), from $0.81 to $22.58.

32. Burkholder 1928.

33. Burkholder letter to Quince Record and Dillon, November 28, 1928.

34. Middle Rio Grande Conservancy District, "Appraisal Data F-10-P5, Serial Number 11N3E 31-6, 1927–1928, Armenta Pedro" (Albuquerque, NM: Archives, MRGCD). The *Rio Grande Drainage Survey*, which was conducted from 1917 to 1918, is on file at the Office of the State Engineer, Santa Fe, New Mexico. For the 1928 reassessment see Middle Rio Grande Conservancy District, "Appraisal Data F-10-P5."

35. Middle Rio Grande Conservancy District, "Classification of Lands for Appraisal of Benefits and Damages," September 24, 1926, revised November 8, 1926, p. 2 (Albuquerque, NM: Archives, MRGCD). See also "Appraisal Data Sheet, MRGCD, Land Classification, Inherent Fertility" (Albuquerque, NM: Archives, MRGCD).

36. *Albuquerque Journal* 1930.

37. W. A. Sutherland, *Memorandum for Judge Helmick in re Petition for Removal of the Conservancy District Board*, June 20, 1930, docket 14157, no. 282; quoted in Orona 1999, 185.

38. Flood Control Act of 1948; *Middle Rio Grande Water Users Association v. Middle Rio Grande Conservancy District*, 57 N.M. 287 (1953).

39. *Albuquerque Journal* 1930.

40. From Forrest (1989). The Taos art colony and a group of independently wealthy individuals from New York—John Collier and Witter Byner among others—invested considerable energy in reinvigorating the New Mexican villages that had fallen on hard times in the 1920s and 1930s. An effort to revamp the Pojoaque-Nambe schools to teach subsistence farming and the old ways, for example, was rejected by the Hispanic community, who instead wanted education for entry into modern life.

41. Calkins 1936.
42. *Hinderlider v. La Plata River and Cherry Creek Ditch Co.*, 291 U.S. 650 (1934); and 304 U.S. 82 (1938).
43. Hill 1983, 163–200.
44. Hall 2002a, 108–29.

Chapter VIII

1. New Mexico Office of the State Engineer, Technical Data, http://www.ose. state.nm.us/PDF/ISC/BasinsPrograms/RioGrande/Tables/RGCompact_Co_ Deliveries_1955.pdf (accessed November 2009).
2. Hill 1983; S. Reynolds and Mutz 1974.
3. Blundell 1980.
4. S. Reynolds 1956, 11.
5. Hall 2002a, 108–10; Jones 2002; McGraw 1980.
6. Hall 2002a, 108–10; King 1998.
7. Chew 1987, 42–48; Workman and Reynolds 1950.
8. Ed Imhoff, interview by G. Emlen Hall, March 10, 2009.
9. McGraw 1980.
10. *State of Texas v. State of New Mexico and MRGCD*, 1951.
11. I. Clark 1987, 388; *Flood Control Act of 1948*, Public Law 858, 80th Cong., 1st sess. H. Doc. 243; 81st Cong., 1st sess.
12. Flood Control Act of 1948; *Middle Rio Grande Water Users Association v. Middle Rio Grande Conservancy District*, 57 N.M. 287 (1953).
13. *Middle Rio Grande Water Users Association v. Middle Rio Grande Conservancy District*.
14. I. Clark 1987, 387–90; H. Doc. 653, 190ff., gives a detailed description of the MRGCD's financial woes in 1951; U.S. Bureau of Reclamation 1951, 10.
15. Welsh 1987, 109–29.
16. Jones 2002; Theis 1953.
17. See chart of accumulated deficits, plate 17.
18. Well hydraulics is the science of how water levels in wells respond to pumping, a subject of critical importance because it is the main tool hydrologists and engineers use to understand the movement of groundwater. New Mexico Tech president E. J. Workman was so impressed with his well hydraulics expertise that he offered Hantush a job heading up an investigation of the aquifers near Roswell. Hantush accepted and moved to New Mexico for 1954–55 (Hall 2002, 105–6).
19. Hantush 1955.
20. Theis 1935.
21. Theis 1940.
22. Theis 1941.
23. Sterrett 2007.
24. Theis 1941.
25. Theis 1953, 19.
26. Ibid., 20. Italics were added by the authors for emphasis.
27. Ibid., 20.
28. The magnitude of Rio Grande losses to groundwater pumping in the 1950s is difficult

to reconstruct. It could possibly have been only a small fraction of the annual compact deficits, but could also have been a substantial proportion. The Office of the State Engineer estimates that, at the time of the 1956 declaration of the Rio Grande Underground Basin, the City of Albuquerque, with its wells, was already depleting the surface flows of the Rio Grande by 18,000 acre-feet per year (Flanigan and Haas 2008, 398–99). The amount that New Mexico underdelivered under the compact was also around 20,000 acre-feet per year.

29. Hall 2002a, 99–100.
30. Jones 2002, 941–53; Hall 2002a, 112–15.
31. Jones 2002, 944.
32. Ibid., 945.
33. Ibid., 948.
34. Shomaker 2007, 1–13.
35. The quote is taken from remarks made by Richard A. Simms at the memorial service for S. E. Reynolds, May 19, 1990, in Santa Fe (possession of G. Emlen Hall); *State of Texas v. States of New Mexico and Colorado and the MRGCD*, Original No. 9, Supreme Court.
36. Comprehensive Plan "for channelization of Rio Grande and rehabilitation of existing MRGCD facilities" approved by the Flood Control Act of 1948. Construction was authorized to begin by the Flood Control Act of 1950 (Public Law 516, 81st Cong., 2nd sess.). See U.S. Bureau of Reclamation (1951, 12).
37. U.S. Bureau of Reclamation 1956, 27.
38. Reisner 1987; Hall 2002a, 115–16.
39. Nabil Shafik (senior engineer, New Mexico Interstate Stream Commission), interview by G. Emlen Hall, April 14, 2009; Steve Hansen (assistant director, Albuquerque office, U.S. Bureau of Reclamation), interview by G. Emlen Hall, May 16, 2009.
40. Bloom and Olson 2008; I. Clark 1987, 508–12; Kelly et al. 2007, 576–96.
41. Kelly et al. 2007, 582, 307n.
42. Carl Slingerland, personal communication with G. Emlen Hall, April 2, 1995.
43. Bloom and Olsen 2008, 17–20; Hall 2002b, 9.
44. Kelly et al. 2007, 586.
45. Welsh 1987, 129–65.
46. Ibid., 109–29, 139–65.
47. See prologue; also Welsh (1987, 144–46) and Kelly et al. (2007, 564–69).
48. Bensfield and Hall 1970.
49. Hansen, interview.
50. New Mexico Office of the State Engineer, Technical Data, http://www.ose. state.nm.us/PDF/ISC/BasinsPrograms/RioGrande/Tables/RGCompact_Co_ Deliveries_1955.pdf (accessed November 2009).
51. Grassel 2002.
52. Tetra Tech 2004, 8–37.

Chapter IX

1. Hester 1987.
2. NASA officially credits the photo to the entire crew because the provenance of a single photo on the mission is difficult to precisely determine, but the agency and

other sleuths acknowledge that Schmitt is the likely photographer; see "Last man on moon makes appearance at JSC lecture series" at http://www.jsc.nasa.gov/jscfeatures/articles/000000674.html (accessed January 30, 2010); and E. Hartwell's *Apollo 17: The Blue Marble*, http://www.ehartwell.com/InfoDabble/Apollo_17:_The_Blue_Marble (accessed January 1, 2010).

3. S. Reynolds's personal letter to G. Emlen Hall, July 5, 1983.
4. L. Warren 1997, 71–126.
5. I. Clark 1987.
6. Bensfield and Hall 1970.
7. New Mexico Office of the State Engineer 1969/1978.
8. *Jicarilla Apache Tribe v. United States*, 601 F.2d 2116 (10th Cir. 1979).
9. Clean Water Act of 1972, 33 U.S.C.A. sec. 1251–1387; *Endangered Species Act of 1973*, 16 U.S.C. sec. 1531–44 (2000).
10. R. Johnson, Haight, and Simpson 1977; A. Johnson 1989.
11. Sublette, Hatch, and Sublette 1990.
12. Robert 2005, 20.
13. C. Crawford et al. 1993.
14. C. Crawford address at the 1994 New Mexico Water Resources Research Institute Annual Conference, as quoted in Robert (2005, 25).
15. Tom Udall, attorney general, and Alletta Belin, assistant attorney general, "Opinion No. 98-01," March 27, 1998 (Santa Fe, NM: Office of the Attorney General, 1998).
16. Ibid., 1998.
17. D. L. Sanders and Stephen R. Farris, *Memorandum to State Engineer Tom Turney from State Engineer Legal Services Division*, January 8, 1998 (Santa Fe, NM: Office of the State Engineer, 1998).
18. Katz 2007, 679.
19. Ibid., 681; *Rio Grande Silvery Minnow v. Keys*, 469 F. Supp. 2nd 973 (D.N.M. 2002).
20. Katz 2007, 683.
21. "Chapter One: Wyatt and the Rio Grande," Earth Guardians, e-mail solicitation to subscribers, December 8, 2008, p. 1.
22. Katz 2007, 683.
23. Ibid., 685–87.
24. *Rio Grande Silvery Minnow et al. v. Bureau of Reclamation et al.*, 601 F.3d 1096 (10th Cir. 2010).
25. Alletta Belin, interview by G. Emlen Hall, Santa Fe, NM, March 24, 2009.
26. New Mexico Office of the State Engineer 2009; Haggerty et al. 2008.
27. For more on constructed wetlands technology, see Gearheart (2006).
28. New Mexico Office of the State Engineer 2009.
29. J. Whitney, R. Leutheuser, R. Barrios, D. Kreiner, J. Whipple, S. Shah, and G. Daves, "Water Management Strategy for the Middle Rio Grande Valley" (unpublished paper, November 14, 1996).
30. Alliance for the Rio Grande Heritage 1998.
31. Water Resources Program and University of New Mexico Law School, introduction to materials for conference on instream flows, May 1994, p. 2; Denise Fort, personal communication with G. Emlen Hall, May, 1994.
32. See http://www.middleriogrande.com (accessed July 14, 2009).
33. Steve Hansen, interview by G. Emlen Hall, April 22, 2009.

34. U.S. Bureau of Reclamation 2005.
35. Steve Hansen, interview by G. Emlen Hall and Mary E. Black, March 18, 2009.
36. Ford, Bacon, and Davis 1928; Calkins 1937.
37. "El Cahete Mine Environmental Impact Statement," *Copar Pumice Co. v. Bosworth*, 502 F. Supp. 2nd 1200 (D.N.M 2007).
38. Kosek 2006, vii–viii; Matthews 2007.
39. Pueblo of Santa Ana website, http://www.santaana.org.
40. Hall 2002a, 60–61; *Pecos River Project: Hearing before the Subcommittee on Irrigation and Reclamation of the Committee on Interior and Insular Affairs*, 84th Cong., 2nd sess. (May 10, 1956), 8–19, Tipton statement.
41. *Phreatophyte Control in the Pecos River Basin: Hearing before the Subcommittee on Irrigation and Reclamation of the Committee on Interior and Insular Affairs*, 88th Cong., 1st sess., May 21, 1963, 36–45 (Reynolds statement).
42. Delaney Hall, "That's Right: Cleaning Up the Bosque with Goats," KUNM radio report, July 23, 2002 (Albuquerque, NM: Archives, KUNM).
43. Hall 2002a, 215.
44. Rayburn 2009.
45. Meyer 1984; Margadant 1989.
46. New Mexico Office of the State Engineer, Interstate Stream Commission 2009, 40.
47. Hall 2008b.
48. Riding 1985, 299–301.
49. DuMars 2007.
50. New Mexico Office of the State Engineer, Interstate Stream Commission 2001, 39–40.
51. Flanigan and Haas 2008, 371–405, 371–73.
52. See Hawley and Haase (1992); and Thorn, McAda, and Kernodle (1993).
53. Kelly et al. 2007, 577–86.
54. Flanigan and Haas 2008, 376.
55. Ibid., 376.
56. Ibid., 402–5.
57. Grassel 2002.
58. Robert 2005.
59. Katz 2007; Gillon 2007, 634–37.
60. Ibid.
61. U.S. Army Corps of Engineers, Albuquerque office, "URGWOM Summary," http://www.spa.usace.army.mil/urgwom/default.asp (accessed January 30, 2009).
62. Flanigan 2007, 521.

Chapter X

1. *U.S. Water News Online*, "Once-Mighty Rio Grande Is Trickle When It Reaches Gulf of Mexico," August 2001, http://www.uswaternews.com/archives/arcsupply/1oncmig8.html (accessed September 1, 2010).
2. Sherman 2010.
3. Intergovernmental Panel on Climate Change 2007a, 10.
4. Diffenbaugh, Giorgi, and Pal 2008.
5. Meko et al. 2007.
6. Grissino-Mayer et al. 2002.
7. Milly et al. 2008.

8. Levi 2008.
9. Wigley and Jones 1985.
10. Milly et al. 2008.
11. Intergovernmental Panel on Climate Change 2007b, 52.
12. U.S. Forest Service 2003.
13. Breshears et al. 2005, 15147.
14. Swetnam 2009.
15. CH2M Hill 2003, 2–7.
16. El Paso Regional Development Corporation, 2009, based on U.S. Bureau of the Census data, http://www.elpasoredco.org/Juarez-Population.aspx (accessed July 2009).
17. El Paso Regional Development Corporation, 2009, based on State of Chihuahua CIES data.
18. Rayburn 2008.
19. Hawley and Haase 1992.
20. Jones 2002, 961–63.
21. Rayburn 2008.
22. Olson 2009.
23. New 2009.
24. Sandia National Laboratories, "Desalination of Saline and Brackish Water Becoming More Affordable," March 19, 2009, http://www.sandia.gov/news/resources/news_releases/desalination-of-saline-and-brackish-water/ (accessed January 23, 2010).
25. M. Hightower, Sandia National Laboratories, quoted in Olson (2008).
26. Olson 2008.
27. Thomson 2008.
28. Reisner 1987, 13.
29. Huddy 2005.
30. City of Santa Fe, *Water Conservation* website, http://www.santafenm.gov/index.aspx?NID=1110 (accessed June 7, 2009).
31. Owen 2008.
32. Ward and Pulido-Velazquez 2008, 18215.
33. Fleming and Hall 2000.
34. Gammage 2005.
35. Lougeay, Brazel, and Hubble 1996.
36. Lovell 2006.
37. "In the Matter of Application of Intel Corporation for Permit to Change Point of Diversion and Place and Purpose of User from Surface to Ground Water in the Rio Grande Underground Basin of the State of New Mexico," Application No. 04246 through 04255 in RG-57125, before the State Engineer, State of New Mexico, December 1999 (Albuquerque, NM: Office of the State Engineer), http://www.ose.state.nm.us/isc_news_IntelAgreement.html (accessed September 15, 2010).
38. S. S. Papadopulos and Associates Inc. 2000, fig. 5–10.
39. Ibid., fig. ES-4.
40. *Middle Rio Grande Regional Water Plan, 2000–2050* (Albuquerque, NM: Middle Rio Grande Water Assembly, 2003).
41. W. M. Turner, "Written Testimony of Dr. William M. Turner, Trustee, Lion's Gate Water," prepared for the House Committee on Resources, Subcommittee on Water and Power, *Hearing on the Silvery Minnow Impact on New Mexico*, held in Belen, NM,

on September 6, 2003, http://www.waterbank.com/Newsletters/nws44.html (accessed July 2009).

42. *Lion's Gate Water v. John D'Antonio, State Engineer for the State of New Mexico*, 2009 NMSC-057, established the authority of the Office of the State Engineer to refuse to process an application to appropriate water on the grounds that there was no water available in any case.

43. Gleick 2008.

44. Millennium Ecosystem Assessment Board 2003.

45. The philosophic basis for ecosystem services was arguably laid down by Daily (1997) and Daily and Ellison (2003).

46. Havstad 2007, 265.

47. See Public Law 110-234, 122 Stat. 923, enacted May 22, 2008, H.R. 2419, also known as the 2008 U.S. Farm Bill, particularly section 2709.

48. K. Smith and Rich 2009.

49. Young 2009.

50. McKay 2005.

51. Flaccus 2007.

52. Ibid.

53. Bloom 2003, 1–24.

Chapter XI

1. Hume 2009.

2. Bernhardt et al. 2005.

3. Seager et al. 2007; Hoerling and Eischeid 2007.

4. S. S. Papadopulos and Associates Inc. 2000, 70.

5. Proverbs 29:18.

6. When Utton (1994) used the phrase "education by catastrophe" in talks and articles, he frequently credited R. Clark (1978, 157).

Bibliography

Abbott, David R. 2003. The politics of decline in Canal System 2. In *Centuries of decline during the Hohokam Classic Period at Pueblo Grande*, ed. D. R. Abbott, 201–27. Tucson: University of Arizona Press.

Adler, Michael A., Todd Van Pool, and Robert D. Leonard. 1996. Ancestral Pueblo aggregation and abandonment in the Southwest. *Journal of World Prehistory* 10 (3): 375–438.

Akins, Nancy J. 1986. *A biocultural approach to human burials from Chaco Canyon, New Mexico*. Reports of the Chaco Center 9. Santa Fe, NM: Division of Cultural Research, National Park Service.

Albuquerque Journal. 1930. City celebrates conservancy work start. March 21.

Allen, C. D., M. Savage, D. A. Falk, K. F. Suckling, T. W. Swetnam, T. Schulke, P. B. Stacey, P. Morgan, M. Hoffman, and J. T. Klingel. 2002. Ecological restoration of southwestern ponderosa pine ecosystems: A broad perspective. *Ecological Applications* 12 (5): 1418–33.

Alliance for the Rio Grande Heritage. 1998. Middle Rio Grande: Water plan for the future. Unpublished paper, June 3.

Anderson, Bertin, and Cameron Barrows. 1998. The debate over tamarisk: The case for wholesale removal. *Ecological Restoration* 16 (2): 135–38.

Anon. 1598/1953. Record of the marches by the army, New Spain to New Mexico (1598). In *Don Juan de Oñate, colonizer of New Mexico: 1595–1628*, ed. G. P. Hammond and A. Rey, 309–28. Albuquerque: University of New Mexico Press.

Anschuetz, Kurt F. 1998. Not waiting for the rain: Integrated systems of water management for intensive agricultural production in north-central New Mexico. PhD thesis, University of Michigan.

Autobee, Robert. 1994. *Rio Grande Project (third draft)*. Bureau of Reclamation History Program. http://www.usbr.gov/projects/Project.jsp (accessed April 16, 2009).

Barrett, Elinore M. 2002. The geography of the Rio Grande pueblos in the seventeenth century. *Ethnohistory* 49 (1): 123–69.

Bartolino, James R., and James C. Cole. 2002. *Ground-water resources of the middle Rio Grande basin.* Water-Resources Circular 1222. Denver, CO: U.S. Geological Survey.

Bastien, Elizabeth M. 2009. Solute budget of the Rio Grande above El Paso, Texas. MS thesis, New Mexico Institute of Mining and Technology.

Bates, J. Leonard. 1957. Fulfilling American democracy: The conservation movement 1907 to 1921. *Mississippi Valley Historical Review* 44:29–57.

Baxter, John O. 1990. *Spanish irrigation in the Taos Valley.* Santa Fe, NM: Office of the State Engineer.

———. 1997. *Dividing New Mexico's Waters, 1700–1912.* Albuquerque: University of New Mexico Press.

———. 2000. Measuring New Mexico's irrigation water: How big is a "surco"? *New Mexico Historical Review* 75 (3): 397–413.

Bensfield, Jim, and G. Emlen Hall. 1970. Cochiti Lake: The making of the seven-day weekend. *New Mexico Review and Legislative Journal* 2 (11): 1–2, 6–8.

Benson, Larry V., and Michael S. Berry. 2009. Climate change and cultural response in the prehistoric American Southwest. *Kiva* 75 (1): 89–119.

Bernhardt, E. S., M. A. Palmer, J. D. Allan, G. Alexander, K. Barnas, S. Brooks, J. Carr, et al. 2005. Synthesizing U.S. river restoration efforts. *Science* 308:636–37.

Betancourt, Julio L., Jeffrey Dean, and Herbert Hull. 1986. Prehistoric long-distance transport of construction beams, Chaco Canyon, New Mexico. *American Antiquity* 51:370–75.

Betancourt, Julio L., and T. R. Van Devender. 1981. Holocene vegetation in Chaco Canyon, New Mexico. *Science* 214:656–58.

Bien, Morris. 1905. Proposed state code of water laws. Paper presented at the Second Conference of Engineers of the Reclamation Service, compiled by F. H. Newell, chief engineer.

Bindeman, Ilya N. 2006. The secrets of supervolcanoes. *Scientific American* 294 (6): 36–43.

Blakey, Ron, and Wayne Ranney. 2008. *Ancient landscapes of the Colorado Plateau.* Grand Canyon, AZ: Grand Canyon Association.

Bloodgood, Dean W. 1921. *Drainage in the Mesilla Valley of New Mexico.* Agriculture Experiment Station Bulletin 129. State College: New Mexico College of Agriculture and Mechanic Arts.

Bloom, David E., David Canning, and Jaypee Sevilla. 2003. *The demographic dividend: A new perspective on the economic consequences of population change.* Santa Monica, CA: Rand Corporation.

Bloom, Paul, and Thomas Olson. 2008. The birth of the San Juan/Chama Diversion Project. In *Materials for course on San Juan/Chama Project,* 1–24. Denver, CO: Continuing Legal Education.

Blundell, William. 1980. Hot spot: In New Mexico, water is valuable resource—and so is water boss. *Wall Street Journal*, April 15.

Bowden, Jocelyn Jean. 1969. Private land claims in the Southwest. MA thesis, Southern Methodist University.

Boyd, Nathan E. 1902. *Statement by Dr. Nathan Boyd in re Elephant Butte (New Mexico) Dam enterprise and General Anson Mills's International (El Paso) Dam scheme: Proposed treaty with the Republic of Mexico, injunction suit, United States v. Rio Grande Dam and Irrigation Company*. Abstracts from official documents and from the decisions of the courts. Washington, D.C.: Judd and Detweiler.

Boyer, Jeffrey L., James L. Moore, Steven A. Lakatos, Nancy J. Akins, C. Dean Wilson, and Eric Blinman. Forthcoming. Remodeling immigration: A northern Rio Grande perspective on depopulation, migration, and donation-side models. In *Collapse and migration in the 13th-Century Puebloan Southwest*, ed. Timothy A. Kohler, Mark D. Varien, and Aaron Wright. Tucson: University of Arizona Press.

Breshears, David D., Neil S. Cobb, Paul M. Rich, Kevin P. Price, Craig D. Allen, Randy G. Balice, William H. Romme, et al. 2005. Regional vegetation die-off in response to global-change-type drought. *Proceedings of the National Academy of Sciences* 102:15144–48.

Brinkley, Douglas. 2009. *The wilderness warrior: Theodore Roosevelt and the crusade for America*. New York: HarperCollins.

Broadbent, Craig D., David S. Brookshire, Don Coursey, and Vincent C. Tidwell. 2010. Creating real-time water leasing market institutions: An integrated economic and hydrological methodology. *Journal of Contemporary Water Research and Education* 144:50–59.

Bryan, Kirk. 1928. Historic evidence on changes in the channel of the Rio Puerco, a tributary of the Rio Grande in New Mexico. *Journal of Geology* 36:265–82.

Bullard, Thomas F., and Stephen G. Wells. 1992. *Hydrology of the middle Rio Grande from Velarde to Elephant Butte Reservoir*. U.S. Fish and Wildlife Service Resource Publication 179. Washington, D.C.: U.S. Fish and Wildlife Service.

Burkholder, Joseph L. 1928. *Report of the chief engineer, Joseph L. Burkholder, submitting a plan for flood control, drainage, and irrigation of the Middle Rio Grande Conservancy project*. Albuquerque, NM: Middle Rio Grande Conservancy District.

Calkins, Hugh G. 1936. *Reconnaissance survey of human dependency on resources in the Rio Grande watershed*. USDA Soil Conservation Service, Region 8, Regional Bulletin 33. Albuquerque, NM: U.S. Department of Agriculture,

———. 1937. *The Santa Cruz Irrigation District, New Mexico*. USDA Soil Conservation Service, Region 8, Regional Bulletin 45; Conservation Economics Survey 18. Albuquerque, NM: U.S. Department of Agriculture.

———. 1941. Man and gullies. *New Mexico Quarterly Review* 11:69–78.

Carlson, Alvar Ward. 1967. Rural settlement patterns in the San Luis Valley: A comparative study. *Colorado Magazine* 44:112–28.

Cather, S. M. 2004. Laramide orogeny in central and northern New Mexico and southern Colorado. In Mack and Giles 2004, 203–48.

CH2M Hill. 2003. Hydrologic effects of the proposed City of Albuquerque Drinking Water Project on the Rio Grande and Rio Chama systems, updated for new conservation and curtailment conditions, October 2003. In *City of Albuquerque Drinking Water Project, environmental impact statement,* appendix L. Albuquerque, NM: City of Albuquerque Public Works Department, Water Resources Strategy Implementation.

Chapin, C. E., W. C. McIntosh, and R. M. Chamberlin. 2004. The Late Eocene-Oligocene peak of Cenozoic volcanism in southwestern New Mexico. In Mack and Giles 2004, 271–93.

Chew, Joe. 1987. *Storms above the desert: Atmospheric research in New Mexico, 1935–1985.* Albuquerque: University of New Mexico Press.

Clark, Ira G. 1987. *Water in New Mexico: A history of its management and use.* Albuquerque: University of New Mexico Press.

Clark, Robert Emmet. 1978. Institutional alternatives for managing groundwater resources: Notes for a proposal. *Natural Resources Journal* 18 (1): 153–62.

Conard, Howard Louis. 1891. *Uncle Dick Wootton.* Chicago: W. E. Dibble.

Connell, S. D. 2004. Geology of the Albuquerque Basin and tectonic development of the Rio Grande rift in north-central New Mexico. In Mack and Giles 2004, 359–88.

Cook, E. R. 2000. Southwestern USA drought index reconstruction. International Tree-Ring Data Bank. IGBP PAGES/World Data Center for Paleoclimatology, Data Contribution Series 2000-053. Boulder, CO: NOAA/NGDC Paleoclimatology Program. http://www.ncdc.noaa.gov/paleo/metadata/noaa-recon-6385.html (accessed September 20, 2010).

Cooperrider, C. K., and B. A. Hendricks. 1937. *Soil erosion and stream flow on range and forest lands of the upper Rio Grande watershed region in relation to land resources and human welfare.* Washington, D.C.: U.S. Department of Agriculture.

Cordell, Linda S. 1984. *Prehistory of the Southwest.* New York: Academic Press.

———. 1994. *Ancient Pueblo peoples.* Montreal: St. Remy Press.

———. 1997. *Archaeology of the Southwest.* 2nd ed. San Diego: Academic Press.

Crawford, Clifford S., Anne C. Cully, Rob Leutheuser, Mark S. Sifuentes, Larry H. White, and James P. Wilber. 1993. *Middle Rio Grande ecosystem: Bosque biological management plan.* Albuquerque, NM: Middle Rio Grande Biological Interagency Team, U.S. Fish and Wildlife Service.

Crawford, Stanley. 1988. *Mayordomo: Chronicle of an acequia in northern New Mexico.* Albuquerque: University of New Mexico Press.

Cummings, Linda Scott, and Thomas E. Moutoux. Pollen analysis. In *Farming through the ages: 3,400 years of agriculture at the Valley Farms site in the northern Tucson Basin*, ed. K. K. Wellman, 275–92. SWCA Cultural Resource Report 98-226. Tucson, AZ: SWCA Inc.

Daily, Gretchen C., ed. 1997. *Nature's services: Societal dependence on natural ecosystems.* Washington, D.C.: Island Press.

Daily, Gretchen C., and Katherine Ellison. 2003. *The new economy of nature: The quest to make conservation profitable.* Washington, D.C.: Island Press.

Damp, Jonathan E., Stephen A. Hall, and Susan J. Smith. 2002. Early irrigation on the Colorado Plateau near Zuni Pueblo, New Mexico. *American Antiquity* 67 (4): 665–76.

Davis, William Watts Hart. 1857. *El Gringo: New Mexico and her people.* New York: Harper and Brothers.

Diamond, Jared. 2005. *Collapse: How societies choose to fail or succeed.* New York: Viking Press.

Diffenbaugh, Noah S., Filippo Giorgi, and Jeremy S. Pal. 2008. Climate change hotspots in the United States. *Geophysical Research Letters* 35, L16709, doi:10.1029/2008GL035075.

Dominguez, Francisco Atanasio. 1956. *The missions of New Mexico, 1776: A description by Fray Francisco Atanasio Dominguez, with other contemporary documents.* Ed. Eleanor Adams and Fray Angelico Chavez. Albuquerque: University of New Mexico Press.

Dortignac, E. J. 1956. *Watershed resources and problems of the upper Rio Grande basin.* Fort Collins, CO: Rocky Mountain Research Station, U.S. Forest Service.

Dozier, Edward P. 1970. *The Pueblo Indians of North America.* Prospect Heights, IL: Waveland Press.

duBuys, William Eno. 1985. *Enchantment and exploitation: Life and times of a New Mexico mountain range.* Albuquerque: University of New Mexico Press.

DuMars, Charles T. 2007. Do we still need adjudications? Some thoughts on the future of water rights adjudications. Paper presented at the American Bar Association 25th Annual Water Law Conference: Changing Values, Changing Conflicts, San Diego, CA, February 23.

Ebright, Malcolm. 1994. *Land grants and lawsuits in northern New Mexico.* Albuquerque: University of New Mexico Press.

Ellis, Florence Hawley. 1970. Irrigation and water works in the Rio Grande valley. In *Water Control Systems Symposium, organized by J. C. Kelley.* 58th Annual Pecos Conference. Santa Fe: Laboratory of Anthropology, Museum of New Mexico.

———. 1979. Summaries of the history of water use and the Tewa culture of the Pojoaque Valley Pueblos. In *State of New Mexico v. R. Lee Aamodt et al.* Santa Fe: Museum of New Mexico.

Ellis, Florence Hawley, and A. E. Dodge. 1989. The spread of Chaco/Mesa Verde/McElmo black-on-white pottery and the possible simultaneous introduction of irrigation into the Rio Grande drainage. *Journal of Anthropological Research* 45:47–52.

Ellis, S. R., G. W. Levings, L. F. Carter, S. F. Richey, and M. J. Radell. 1993. Rio Grande valley, Colorado, New Mexico, and Texas. *Water Resources Bulletin* 29: 617–46.

English, Nathan, Julio Betancourt, Jeffrey Dean, and Jay Quade. 2001. Strontium isotopes reveal distinct sources of architectural timber in Chaco Canyon, New Mexico. *Proceedings of the National Academy of Sciences, USA* 98:11891–96.

Ferrari, Ronald L. 2008. *Elephant Butte Reservoir 2008 sedimentation survey*. U.S. Bureau of Reclamation Technical Report SRH-2008-4. Denver, CO: U.S. Bureau of Reclamation.

Fish, Suzanne K. 1998. Cultural pollen. In *Archaeological investigations of early village sites in the middle Santa Cruz Valley: Analyses and synthesis*, pt. 1, ed. J. B. Mabry, 149–63. Anthropological Papers 19. Tucson, AZ: Center for Desert Archaeology.

Flaccus, Gillian. 2007. The desert oasis that water built is drying up. *Albuquerque Journal*, December 21.

Flanigan, Kevin G. 2007. Surface water management: Working within the legal framework. *Natural Resources Journal* 47 (3): 515–23.

Flanigan, Kevin G., and Amy L. Haas. 2008. The impact of full beneficial use of San Juan–Chama Project water by the City of Albuquerque on New Mexico's Rio Grande Compact obligation. *Natural Resources Journal* 48 (2): 371–406.

Fleming, William M., and G. Emlen Hall. 2000. Water conservation incentives for New Mexico: Policy and legislative alternatives. *Natural Resources Journal* 40 (1): 69–91.

Follett, William W. 1896. *A study of the use of water for irrigation on the Rio Grande del Norte above Ft. Quitman, Texas (November, 1896), proceedings of the International Boundary Commission*. Santa Fe, NM: Office of the State Engineer.

Force, Eric, R. Gwinn Vivian, Thomas Windes, and Jeffrey Dean. 1990. *Relation of "Bonito" paleo-channel and base-level variations to Anasazi occupation, Chaco Canyon, New Mexico*. Arizona State Museum Archaeological Series 194. Tucson: University of Arizona Press.

Ford, Bacon and Davis, Engineers. 1928. *Report, Santa Cruz Irrigation District, Santa Fe and Rio Arriba Counties, New Mexico, August 17, 1928*. Espanola, NM: Santa Cruz Irrigation District.

Forrest, Suzanne. 1989. *The preservation of the village: New Mexico's Hispanics and the New Deal*. Albuquerque: University of New Mexico Press.

Fox, M. L. 1928. The public forum: Press needs less politics. *Albuquerque Tribune*, January 24.

Frazier, Arthur H., and Wilbur Heckler. 1972. *Embudo, New Mexico, birthplace of systematic gaging.* U.S. Geological Survey Professional Paper 78. Washington, D.C.: GPO.

Friedman, Richard A., John R. Stein, and Taft Blackhorse Jr. 2003. A study of a pre-Columbian irrigation system at Newcomb, New Mexico. *Journal of GIS in Archaeology* 1:4–10.

Gammage, Grady, Jr. 2005. Water, growth, and the future of agriculture. *Southwest Hydrology* 4 (4): 28–29.

Gault, H. J. 1914. *Report on surveys at San Marcial, New Mexico, for the purpose of securing data from which to determine the possible future effects of Elephant Butte Reservoir.* Las Cruces, NM: U.S. Reclamation Service.

———. 1923. *Report on the Rio Grande Reclamation Project.* Denver, CO: U.S. Reclamation Service.

Gearheart, Robert A. 2006. Constructed wetlands for natural water treatment. *Southwest Hydrology* 5 (1): 16–17.

Gellis, Allen. 1992. Decreasing trends of suspended sediment concentrations at selected streamflow stations in New Mexico. Paper presented at the 36th Annual New Mexico Water Conference, Las Cruces. New Mexico Water Resources Research Institute Report 265. Las Cruces: New Mexico Water Resources Research Institute.

———. 2006. *History of streamflow and suspended-sediment collection in the Rio Puerco Basin, New Mexico.* U.S. Geological Survey On-Line Report. http://esp.cr.usg.gov/rio-puerco/erosion/streamflows.html (accessed January 2010).

Gillon, Kara. 2007. An environmental pool for the Rio Grande. *Natural Resources Journal* 47 (3): 615–38.

Gleick, Peter H. 1996. Basic water requirements for human activities: Meeting basic needs. *Water International* 21:83–92.

———. 2007. Human right to water. *Water Policy* 1 (5): 487–503. http://www.pacinst.org/reports/human_right_may_07.pdf.

Glenn, Edward P., and Pamela L. Nagler. 2005. Comparative ecophysiology of *Tamarix ramosissima* and native trees in western U.S. riparian zones. *Journal of Arid Environments* 61:419–46.

Glenn, Edward P., Pamela L. Nagler, and Jeffrey E. Lovich. 2009. The surprising value of saltcedar. *Southwest Hydrology* 8 (3): 10–11.

Glick, Thomas F. 1970. *Irrigation and society in medieval Valencia.* Cambridge, MA: Belknap Press.

Granjon, Henry. 1986. *Along the Rio Grande: A pastoral visit to southwest New Mexico in 1902.* Ed. Michael Romero Taylor. Albuquerque: University of New Mexico Press.

Grassel, Kathy. 2002. *Taking out the jacks: Issues of jetty jack removal in bosque and river restoration planning.* Water Resources Program Report WRP-6. Albuquerque: University of New Mexico.

Graybill, Donald A., David A. Gregory, Gary S. Funkhouser, and Fred L. Nials. 2002. Long-term streamflow reconstructions, river channel morphology, and aboriginal irrigation systems along the Salt and Gila rivers. In *Environmental change and human adaptation in the American Southwest*, ed. David E. Doyel and Jeffrey S. Dean, 69–123. Salt Lake City: University of Utah Press.

Gregg, Josiah. 2001. *The commerce of the prairies: Life on the Great Plains in the 1830s and 1840s*. Santa Barbara, CA: Narrative Press.

Grissino-Mayer, Henri D., Christopher H. Baisan, Kiyomi A. Morino, and Thomas W. Swetnam. 2002. *Multi-century trends in past climate for the middle Rio Grande basin, AD 622–1992. Final Report*. Albuquerque, NM: U.S. Forest Service.

Gumerman, George J., and Jeffrey S. Dean. 1989. Prehistoric cooperation and competition in the western Anasazi area. In *Dynamics of Southwest prehistory*, ed. Linda Cordell and G. J. Gumerman. Washington, D.C.: Smithsonian Institution Press.

Haggerty, Grace M., Douglas Tave, Rolf Schmidt-Petersen, and John Stomp. 2008. Raising endangered fish in New Mexico. *Southwest Hydrology* 4:20–21.

Hall, G. Emlen. 1983. Water: New Mexico's delicate balance. *New Mexico Magazine* 61 (5): 15–20.

———. 1984. *Four leagues of Pecos: A legal history of the Pecos Grant, 1800–1934*. Albuquerque: University of New Mexico Press.

———. 1987. Pueblo grant labyrinth. In *Land, water, and culture: New perspectives on Hispanic land grants*, ed. Charles L. Briggs and John R. Van Ness, 67–138. Albuquerque: University of New Mexico Press.

———. 1988. Land litigation and the idea of New Mexico progress. *Journal of the West* 27:48–58.

———. 1993. Community land grants and the Forest Service as watershed managers: The example of Santo Domingo de Cundiyo. Paper presented at Making Sustainability Operational: 4th Mexico/U.S. Symposium, Santa Fe, NM, April 19–23. General Technical Report RM-240. Fort Collins, CO: Rocky Mountain Research Station, U.S. Forest Service.

———. 2000. Tularosa and the dismantling of New Mexico community ditches. *New Mexico Historical Review* 75 (1): 76–106.

———. 2002a. *High and dry: The Texas–New Mexico struggle for the Pecos River*. Albuquerque: University of New Mexico Press.

———. 2002b. How the Gila got its spots. Paper delivered under the aegis of the New Mexico Humanities Council, Silver City, NM, March 24.

———. 2008a. The first 100 years of the New Mexico Water Code. *Natural Resources Journal* 48 (2): 245–48.

———. 2008b. Adjudication nightmares. Paper presented at the Western State Conference on Adjudications, Santa Fe, NM, October 2008. On file at the Office of the State Engineer, Santa Fe.

Hantush, Mahdi Salih. 1955. Discussion of "River depletion resulting from pumping a well near the river" by G. G. Balmer and R. E. Glover. *Transactions–American Geophysical Union* 36 (2): 345–46.

Harden, Paul. 2006. Socorro County floods. *El Defensor Chieftain*, September 6.

Harris, Linda G. 1996. The developers: Controlling the lower Rio Grande 1890–1980. Paper presented at the 40th Water Resources Research Institute Annual Water Conference, Socorro, NM, October 26–27, 1995. *Water Resources Research Institute/40th Proceedings* 297:7–12.

Harroun, Philip E. 1899. Present condition of irrigation and water supply in New Mexico. In *Commission of irrigation and water rights, report to territorial governor Otero, December 15, 1898*, 23–75. Santa Fe: New Mexico Printing Company.

Havstad, K. M., D. C. Peters, R. Skaggs, J. Brown, B. T. Bestelmeyer, E. L. Fredrickson, J. E. Herrick, and J. Wright. 2007. Ecological services to and from rangelands of the United States. *Ecological Economics* 64:261–68.

Hawley, J. W. 2005. Five million years of landscape evolution in New Mexico: An overview based on two centuries of geomorphic conceptual-model development. In *New Mexico's Ice Ages*, ed. S. G. Lucas, G. S. Morgan, and K. E. Zeigler, 9–94. Albuquerque: New Mexico Museum of Natural History and Science.

Hawley, J. W., and C. S. Haase. 1992. Hydrogeologic framework of the northern Albuquerque basin. Open File Report 387. Socorro: New Mexico Bureau of Geology and Mineral Resources.

Hayden, F. V. 1869. *Preliminary field report of the United States Geological Survey of Colorado and New Mexico*. Washington, D.C.: GPO.

Haynes, C. Vance, Jr. 1991. Geoarchaeological and paleohydrological evidence for a Clovis-age drought in North America and its bearing on extinction. *Quaternary Research* 35 (3): 438–50.

———. 2008. Younger Dryas "black mats" and the Rancholabrean termination in North America. *Proceedings of the National Academy of Sciences, USA* 105 (18): 6520–25.

Hays, Samuel P. 1959. *Conservation and the gospel of efficiency*. Cambridge, MA: Harvard University Press.

———. 1975. *Conservation and the gospel of efficiency: The Progressive conservation movement, 1890–1920*. New York: Atheneum.

Hedke, C. R. 1925. *A report on the irrigation development and water supply of the middle Rio Grande valley, NM, as it relates to the Rio Grande Compact*. Santa Fe, NM: Rio Grande Valley Survey Commission, Office of the State Engineer.

Hester, Nolan. 1987. The water boss. *Impact: The Albuquerque Journal Magazine*, May 5, 4–9.

Hill, Raymond A. 1974. Development of the Rio Grande Compact of 1938. *Natural Resources Journal* 14 (2): 163–98.

Hoerling, Martin, and Jon Eischeid. 2007. Past peak water in the Southwest. *Southwest Hydrology* 6 (1): 18–19, 35.

Hogan, J. F., F. M. Phillips, S. K. Mills, J. M. H. Hendrickx, J. Ruiz, J. T. Chesley, and Y. Asmerom. 2007. Geological origins of salinization in a semiarid river: The role of sedimentary brines. *Geology* 35:1063–66.

Howland, Harold. 1921. *Theodore Roosevelt and his times: A chronicle of the Progressive movement*. New Haven, CT: Yale University Press.

Huckell, Bruce B. 1995. *Of marshes and maize: Preceramic agricultural settlements in the Cienega Valley, southeastern Arizona*. Anthropological Papers 59. Tucson: University of Arizona Press.

———. 1998. Alluvial stratigraphy of the Santa Cruz Bend reach. In *Archaeological investigations of early village sites in the middle Santa Cruz Valley: Analyses and synthesis*, ed. J. B. Mabry, 31–56. Anthropological Papers 19. Tucson, AZ: Center for Desert Archaeology.

Huckleberry, Gary A., and Brian R. Billman. 1998. Floodwater farming, discontinuous ephemeral streams, and puebloan abandonment in southwestern Colorado. *American Antiquity* 63 (4): 595–614.

Huddy, J. T. 2005. Santa Fe City Council kills Estancia water plan. *Albuquerque Journal*, January 27.

Hume, Bill. 2009. Concerns over water consumption are valid. *Albuquerque Journal*, November 5, 2009.

Intergovernmental Panel on Climate Change. 2007a. *Climate change 2007: Synthesis report, summary for policymakers*. Geneva, Switzerland: Intergovernmental Panel on Climate Change.

———. 2007b. *Climate change 2007 synthesis report: Contribution of working groups I, II, and III to the fourth assessment report of the Intergovernmental Panel on Climate Change*. Ed. P. Core writing team, R. K. Pachauri, and A. Reisinger. Geneva, Switzerland: Intergovernmental Panel on Climate Change.

Jenkins, Myra Ellen, and John O. Baxter. 1986. *Settlement and irrigation in the Santa Cruz-Quemado watershed*. Santa Fe, NM: Office of the State Engineer.

Johnson, Aubrey S. 1989. The thin green line: Riparian corridors and endangered species in Arizona and New Mexico. In *In defense of wildlife: Preserving communities and corridors*, ed. G. Mackintosh, 35–46. Washington, D.C.: Defenders of Wildlife.

Johnson, R. R., L. T. Haight, and J. M. Simpson. 1977. Endangered species vs. endangered habitats: A concept. Paper presented at Importance, Preservation, and Management of Riparian Habitat: A Symposium, Fort Collins, CO. General Technical Report RM-43. Fort Collins, CO: Rocky Mountain Research Station, U.S. Forest Service.

Jones, Celina A. 2002. The administration of the Rio Grande basin, 1956–2003. *Natural Resources Journal* 42 (4): 939–53.

Kaplan, Lawrence. 1965. Archaeology and domestication in American *Phaseolus* (beans). *Economic Botany* 19 (4): 358–68.

Karlstrom, K. E., J. M. Amato, M. L. Williams, M. T. Heizler, C. A. Shaw, A. S. Read, and P. Bauer. 2004. Proterozoic tectonic evolution of the New Mexico region: A synthesis. In Mack and Giles 2004, 1–34.

Katz, Laura. 2007. History of the minnow litigation and its implications for the future of reservoir operations on the Rio Grande. *Natural Resources Journal* 47 (3): 675–92.

Kelly, Susan, Iris Augusten, Joshua Mann, and Laura Katz. 2007. History of the Rio Grande reservoirs in New Mexico: Legislation and litigation. *Natural Resources Journal* 47 (3): 525–613.

Kennett, D. J., J. P. Kennett, A. West, C. Mercer, S. S. Que Hee, L. Bement, T. E. Bunch, M. Sellers, and W. S. Wolbach. 2009. Nanodiamonds in the younger Dryas boundary sediment layer. *Science* 323 (5910): 94.

Kerr, Richard A. 2010. Mammoth-killer impact flunks out. *Science* 329 (5996): 1140–41.

King, Bruce. 1998. *Cowboy in the roundhouse: A political life*. Santa Fe, NM: Sunstone Press.

Kosek, Jake. 2006. *Understories: The political life of forests in northern New Mexico*. Durham, NC: Duke University Press.

Kurc, Shirley A., and Eric E. Small. 2004. Dynamics of evapotranspiration in semiarid grassland and shrubland ecosystems during the summer monsoon season, central New Mexico. *Water Resources Research* 40, W09305, doi:10.1029/2004WR003068.

Lacey, Heather. 2006. Quantification and characterization of chloride sources in the Rio Grande. MS thesis, New Mexico Institute of Mining and Technology.

Lakatos, Steven A. 2007. Cultural continuity and the development of integrative architecture in the northern Rio Grande valley of New Mexico, A.D. 600–1200. *KIVA* 73 (1): 31–66.

Lange, Charles H. 1959. *Cochiti: A New Mexico pueblo, past and present*. Austin: University of Texas Press.

Lawton, H. W., and P. J. Wilke. 1979. Ancient agricultural systems in dry regions. In *Agriculture in semi-arid regions*, ed. A. E. Hall, G. H. Cannell, and H. W. Lawton, 1–44. Berlin: Springer-Verlag.

Lekson, Stephen H. 2006. Chaco matters: An introduction. In *The archaeology of Chaco Canyon: An eleventh-century pueblo regional center*, 3–44. Santa Fe, NM: School of American Research Press.

Leopold, Luna B., and V. A. Koelzer. 1947. Science against silt. *Reclamation Era* (February): 39–41.

Lester, Paul. 1977. History of the Elephant Butte Irrigation District. MA thesis, New Mexico State University, Las Cruces.

Lipe, William D. 2006. Notes from the north. In *The archaeology of Chaco Canyon*, ed. S. H. Lekson. Santa Fe, NM: School of American Research Press.

Lippincott, J. B. 1939. Southwestern border water problems. *Journal of the American Water Works Association* 31:1–29.

Lite, Sharon J., and Juliet C. Stromberg. 2005. Surface water and ground-water thresholds for maintaining *Populus–Salix* forests, San Pedro River, Arizona. *Biological Conservation* 125:153–67.

Littlefield, Douglas R. 1987. Interstate water conflicts, compromises, and compacts: The Rio Grande, 1880–1938. PhD thesis, University of California, Los Angeles.

———. 2008. *Conflict on the Rio Grande: Water and law, 1879–1939.* Norman: University of Oklahoma Press.

Lougeay, Ray, Anthony Brazel, and Mark Hubble. 1996. Monitoring intraurban temperature patterns and associated land cover in Phoenix, Arizona, using Landsat thermal data. *Geocarto International* 11 (4): 79–90.

Love, David W. 1997. Implications for models of arroyo entrenchment and distribution of archaeological sites in the middle Rio Puerco. In *Layers of time*, vol. 23, ed. M. S. Duran and D. T. Kirkpatrick, 69–84. Albuquerque: Archaeological Society of New Mexico.

Lovell, Margaret. 2006. The high-tech way to use less water. *Innovation* 2 (6). http://www.innovation-america.org/high-tech-way-use-less-water.

Mabry, Jonathan B. 2005. Changing knowledge and ideas about the first farmers in Southeastern Arizona. In *Current perspectives on the Late Archaic across the borderlands: From foraging to farming*, ed. B. J. Vierra, 41–83. Austin: University of Texas Press.

MacCameron, Robert. 1994. Environmental change in colonial New Mexico. *Environmental History Review* (Summer): 17–39.

Machette, M. N., D. W. Marchetti, and R. A. Thompson. 2007. Ancient Lake Alamosa and the Pliocene to Middle Pleistocene evolution of the Rio Grande. In *Rocky Mountain Section Friends of the Pleistocene Field Trip—Quaternary geology of the San Luis Basin of Colorado and New Mexico, September 7–9, 2007*, 157–67. U.S. Geological Survey Open-File Report 2007-1193. Reston, VA: U.S. Geological Survey.

Mack, G. H. 2004a. The Cambro-Ordovician bliss and lower Ordovician El Paso formations, southwestern New Mexico and west Texas. In Mack and Giles 2004, 35–44.

———. 2004b. Middle and late Cenozoic crustal extension, sedimentation, and volcanism in the southern Rio Grande rift, basin and range, and southern transition zone of southwestern New Mexico. In Mack and Giles 2004, 389–406.

Mack, G. H., and K. A. Giles, eds. 2004. *The geology of New Mexico.* Socorro: New Mexico Geological Society.

Malone, D., ed. 1934. *Dictionary of American biography.* New York: Scribner's.

Margadant, Guillermo F. 1989. *Memorandum on water rights of Indian communities in New Mexico limited to the new Spanish and Mexican periods.* Santa Fe, NM: Office of the State Engineer.

Marshall, James. 1945. *Santa Fe: The railroad that built an empire*. New York: Random House.

Marshall, Michael P., and Henry J. Walt. 1984. *Rio Abajo: Prehistory and history of a Rio Grande province*. Santa Fe: New Mexico Historic Preservation Division.

Martin, Paul S. 2005. *Twilight of the mammoths: Ice Age extinctions and the rewilding of America*. Berkeley: University of California Press.

Matthess, Georg. 1982. *The properties of groundwater*. New York: John Wiley.

Matthews, Kay. 2007. Acequia parciantes turn out to protect water. *La Jicarita News*, January 2007.

McGraw, Kate. 1980. Steve Reynolds is nobody's sweetheart. *New Mexican*, November 10.

McKay, Jennifer. 2005. Water institutional reforms in Australia. *Water Policy* 7 (1): 35–52.

McKenna, Peter J. 1984. *The architecture and material culture of 29S/1360*. Reports of the Chaco Center 7. Albuquerque, NM: Division of Cultural Research, National Park Service.

Meko, D., C. A. Woodhouse, C. A. Baisan, T. Knight, J. J. Lukas, M. K. Hughes, and M. W. Salzer. 2007. Medieval drought in the upper Colorado River basin. *Geophysical Research Letters* 34, doi:10.1029/2007GL029988.

Merryman, John Henry. 1985. *The civil law tradition: An introduction to the legal systems of Western Europe and Latin America*. 2nd ed. Stanford, CA: Stanford University Press.

Meyer, Michael C. 1984. *Water in the Hispanic Southwest: A social and legal history, 1550–1850*. Tucson: University of Arizona Press.

Middle Rio Grande Endangered Species Collaborative Program. *MRGESCP*. http://www.middleriogrande.com (accessed July 14, 2009).

Millennium Ecosystem Assessment Board. 2003. *Ecosystems and human well-being: A framework for assessment*. Washington, D.C.: Island Press.

———. 2005. *Ecosystems and human well-being: Wetlands and water synthesis*. Washington, D.C.: World Resources Institute.

Miller, Calvin F., and David A. Wark. 2008. Supervolcanoes and their explosive supereruptions. *Elements* 4 (1): 11–16.

Mills, Barbara. 2004. Key debates in Chacoan archaeology. In *In search of Chaco: New approaches to an archaeological enigma*, ed. D. Noble, 123–30. Santa Fe, NM: School of American Research Press.

Mills, Suzanne K. 2004. Quantifying salinization of the Rio Grande using environmental tracers. MS thesis, New Mexico Institute of Mining and Technology.

Milly, P. C. D., Julio Betancourt, Malin Falkenmark, Robert M. Hirsch, Zbigniew W. Kundzewicz, Dennis P. Lettermaier, and Ronald J. Stouffer. 2008. Stationarity is dead: Whither water management? *Science* 319 (5863): 573–74.

Moore, S. J., R. L. Bassett, B. Liu, C. P. Wolf, and D. Doremus. 2008. Geochemical tracers to evaluate hydrogeologic controls on river salinization. *Ground Water* 46:489–501.

Murphy, Edward Charles. 1904. Accuracy of stream measurements. U.S. Geological Survey Water Supply Paper 95. Reston, VA: U.S. Geological Survey.

Musselman, Keith, Noah P. Molotch, and Paul D. Brooks. 2008. Quantifying the effects of vegetation on snow accumulation and ablation in a mid-latitude sub-alpine forest. *Hydrological Processes* 22 (15): 2767–76.

Myrick, David F. 1990. *New Mexico's railroads: An historical summary.* Rev. ed. Albuquerque: University of New Mexico Press.

Nagler, Pamela L., Kiyomi Morino, Kamel Didan, Joseph Erker, John Osterberg, Kevin R. Hultine, and Edward P. Glenn. 2008. Wide-area estimates of saltcedar (*Tamarix* spp.) evapotranspiration in the lower Colorado River measured by heat balance and remote sensing methods. *Ecohydrology* 2 (1): 18–33.

Natural Resources Committee. 1938. *Regional planning, part VI: The Rio Grande joint investigation in the upper Rio Grande basin in Colorado, New Mexico, and Texas, 1936–1937, part I: Text, part II: Maps.* Washington, D.C.: GPO.

Newell, Frederick Haynes. 1893. *Water supply for irrigation: Extract from the 13th Annual Report of the Director, 1891–1892.* Washington, D.C.: GPO.

New Mexico Interstate Stream Commission. 2010. Rio Grande Compact accounting: Compact annual compilation tables, 1940–2005. Santa Fe, NM: Office of the State Engineer. http://www.ose.state.nm.us/isc_rio_grande_tech_compact_tables.html, Santa Fe (accessed January 29, 2010).

New Mexico Office of the State Engineer. 1969/1978. *A roster, by county, of organizations concerned with surface water irrigation in New Mexico. Special report, New Mexico State Engineer Office, prepared in cooperation with the U.S. Soil Conservation Service and the Agricultural Stabilization and Conservation Service.* Rev. ed. Santa Fe, NM: Office of the State Engineer.

———. 2009. Grand opening celebration. *Water Wise Community Brief* 6 (1): 2. Santa Fe, NM: Office of the State Engineer.

New Mexico Office of the State Engineer, Interstate Stream Commission. 2001. *Bi-annual report 1999–2000.* Santa Fe, NM: Office of the State Engineer.

———. 2007. *Bi-annual report 2005–2006.* Santa Fe, NM: Office of the State Engineer.

———. 2009. *Bi-annual report 2007–2008.* Santa Fe, NM: Office of the State Engineer.

New regulations for deep water in New Mexico. 2009. *Southwest Hydrology* 8 (4): 14.

Nials, Fred L. 2008. Canal geomorphologies. In *Las Capas: Early irrigation and sedentism in a southwestern floodplain,* ed. J. B. Mabry, 149–68. Tucson, AZ: Center for Desert Archaeology.

Noble, David Grant, ed. 2004. *New light on Chaco Canyon.* Santa Fe, NM: School of American Research.

Olson, S. 2008. Brackish water is expensive to clean. *Albuquerque Journal*, December 29.

———. 2009. Lion's Gate after N.M.'s salty water. *Albuquerque Journal*, March 21.

Orona, Kenneth M. 1999. River of culture, river of power: Identity, modernism, and contest in the middle Rio Grande valley, 1848–1947. PhD thesis, Yale University.

O'Sullivan, John L. 1839. The great nation of futurity. *United States Democratic Review* 6 (23): 426–30.

Parmenter, Robert R. 2009. Applying hydrology to land management on the Valles Caldera National Preserve. *Southwest Hydrology* 8 (2): 22–23.

Pazzaglia, F. J., and J. W. Hawley. 2004. Neogene (rift flank) and Quaternary geology and geomorphology. In Mack and Giles 2004, 407–37.

Pecos, Regis. 2007. The history of the Cochiti Lake from the Pueblo perspective. *Natural Resources Journal* 47 (3): 639–52.

Phillips, Bruce G. 2000. Archaeobotany. In *Archaeological investigations at AZ V:13:201, Town of Kearny, Pinal County, Arizona*, comp. C. V. Clark, 5-1 to 5-30. Cultural Resources Report 114. Tempe, AZ: Archaeological Consulting Services.

Phillips, Frederick M., James F. Hogan, Suzanne Mills, and Jan M. H. Hendrickx. 2003. Environmental tracers applied to quantifying causes of salinity in arid-region rivers: Preliminary results from the Rio Grande, southwestern USA. In *Water resource perspectives: Evaluation, management, and policy*, ed. A. S. Alarshan and W. W. Wood, 327–34. Amsterdam: Elsevier.

Pinchot, Gifford. 1947. *Breaking new ground*. New York: Harcourt Brace Jovanovich.

Pisani, Donald J. 1992. *To reclaim a divided West: Water, law and public policy, 1848–1902*. Albuquerque: University of New Mexico Press.

———. 2002. A tale of two commissioners: Frederick H. Newell and Floyd Dominy. Paper presented at History of the Bureau of Reclamation: A Symposium. Las Vegas, NV, June 18.

Plog, Fred, and Amy Elizabeth Gray. 1997. *Ancient peoples of the American Southwest*. New York: Thames and Hudson.

Poore, Henry R. 1894. Condition of 16 New Mexico Indian pueblos, 1890. In *Report on Indians taxed and Indians not taxed in the United States, at the Eleventh Census, 1890*, 396–446. Washington, D.C.: GPO.

Powell, John Wesley. 1879. *Report on the lands of the arid region of the United States, with a more detailed account of the lands of Utah*. Washington, D.C.: GPO.

Rayburn, Rosalie. 2008. Sandoval aquifer tests show huge supply of briny water. *Albuquerque Journal*, November 1.

———. 2009. Casino expansion back on to-do list. *Albuquerque Journal*, July 20.

Reisner, Marc. 1987. *Cadillac desert: The American West and its disappearing water*. New York: Penguin Books.

Reynolds, Amanda C., Julio L. Betancourt, Jay Quade, P. Jonathan Patchett, Jeffrey S. Dean, and John Stein. 2005. 87Sr/86Sr sourcing of ponderosa pine used in Anasazi great house construction at Chaco Canyon, New Mexico. *Journal of Archaeological Science* 32 (7): 1061–75.

Reynolds, R. V., and A. H. Pierson. 1942. *Fuel wood use in the United States 1630–1930*. U.S. Department of Agriculture Circular 641. Washington, D.C.: U.S. Department of Agriculture.

Reynolds, S. E. 1956. New Mexico's water problems. Address given at New Mexico State University, Las Cruces, November 10. In *Speeches of State Engineer Reynolds*. Unpublished typescript in Office of the State Engineer, Santa Fe.

Reynolds, S. E., and Philip B. Mutz. 1974. Water deliveries under the Rio Grande Compact. *Natural Resources Journal* 14 (2): 201–5.

Rich, John L. 1911. Recent stream trenching in the semi-arid portion of southwestern New Mexico, a result of removal of vegetation cover. *American Journal of Science, Fourth Series* 32:237–45.

Riding, Alan. 1985. *Distant neighbors: A portrait of the Mexicans*. New York: Knopf.

Riley, Carroll L. 1995. *Rio del Norte: People of the upper Rio Grande from earliest times to the Pueblo revolt*. Salt Lake City: University of Utah Press.

Rinehart, Alex J., Enrique R. Vivoni, and Paul D. Brooks. 2008. Effects of vegetation, albedo, and solar radiation sheltering on the distribution of snow in the Valles Caldera, New Mexico. *Ecohydrology* 1:253–70.

Rio Grande Project, Central Board of Review. 1917. Report of the Central Board of Review on the Rio Grande Project, New Mexico–Texas, February 1917. *Reclamation Record* 82 (6): 82–86.

———. 1951. *Project History, Middle Rio Grande Project*. Albuquerque, NM: U.S. Bureau of Reclamation.

Ritter, Dale F., R. Craig Kochel, and Jerry Russell Miller. 2002. *Process geomorphology*. 4th ed. Long Grove, IL: Waveland Press.

Robert, Lisa. 2005. *Middle Rio Grande ecosystem bosque biological management plan, the first decade: A review and update*. Albuquerque, NM: Middle Rio Grande Bosque Initiative and Bosque Improvement Group.

Rodriguez, Sylvia. 2006. *Acequia: Water sharing, sanctity and place*. Santa Fe, NM: School for Advanced Research Press.

Roosevelt, Theodore. 1908. *Proceedings of a conference of governors in the White House, Washington, D.C., May 13–15, 1908*. Washington, D.C.: GPO.

Rowley, William D. 2006. *The Bureau of Reclamation: Origins and growth to 1945*. Denver, CO: U.S. Bureau of Reclamation.

Sandvig, Renee M., and Fred M. Phillips. 2006. Ecohydrological controls on soil moisture fluxes in arid to semiarid vadose zones. *Water Resources Research* 42, W08422, doi:10.1029/2005WR004644.

Sargeant, Kathryn, and Mary Davis. 1986. *Shining river, precious land: An oral history of Albuquerque's North Valley*. Albuquerque, NM: Albuquerque Museum.

Schönfeld La Mar, Barbel Hannelore. 1984. Water and land in the Mesilla Valley, New Mexico: Reclamation and its effects on property ownership and agricultural land use. PhD thesis, University of Oregon.

Scurlock, Dan. 1998. *From the Rio to the Sierra: An environmental history of the middle Rio Grande basin.* General Technical Report RMRS-GTR-5. Fort Collins, CO: Rocky Mountain Research Station, U.S. Forest Service.

Seager, R., M. Ting, I. Held, Y. Kushnir, J. Lu, G. Vecchi, H. P. Huang, et al. 2007. Model projections of an imminent transition to a more arid climate in southwestern North America. *Science* 316 (5828): 1181–84.

Sebastian, Lynne. 1992. *The Chaco Anasazi: Sociopolitical evolution in the prehistoric Southwest.* Cambridge: Cambridge University Press.

———. 2006. The Chaco synthesis. In *The archaeology of Chaco Canyon: An eleventh-century Pueblo regional center,* ed. S. H. Lekson, 393–422. Santa Fe, NM: School of American Research Press.

Shafike, Nabil, Salim Bawazir, and James Cleverly. 2007. Native versus invasive plant water use in the middle Rio Grande basin. *Southwest Hydrology* 6 (6): 28–29.

Sherman, Christopher. 2010. Rio Grande flood called unique in timing, scale. *Brownsville Herald,* July 25.

Shomaker, John. 2007. What shall we do with all of this groundwater? *Natural Resources Journal* 47 (4): 781–91.

Siebenthal, C. E. 1910. *Geology and water resources of the San Luis Valley, Colorado.* U.S. Geological Survey Water Supply Report 240. Reston, VA: U.S. Geological Survey.

Simmons, Marc. 1968. *Spanish government in New Mexico.* Albuquerque: University of New Mexico Press.

———. 1969. Settlement patterns and village plans in colonial New Mexico. *Journal of the West* 8 (18): 7–21.

———. 1972. Spanish irrigation practices in New Mexico. *New Mexico Historical Review* 47:135–50.

———. 2003. *Hispanic Albuquerque, 1706–1846.* Albuquerque: University of New Mexico Press.

Simmons, Virginia McConnell. 1979. *The San Luis Valley: Land of the six-armed cross.* Boulder, CO: Pruett.

Simpich, Frederick. 1920. Along our side of the Mexican border. *National Geographic* 38:61–80.

Smith, G. A. 2004. Middle to Late Cenozoic development of the Rio Grande rift and adjacent regions in northern New Mexico. In Mack and Giles 2004, 311–58.

Smith, Ken, and Susan Rich. 2009. Restoration in New Mexico watersheds: The uplands. *Southwest Hydrology* 8 (2): 30, 32–33.

Solstice Project. 1999. *The mystery of Chaco Canyon.* DVD. Produced by the Solstice Project. Directed by Anna Sofaer. Written by Anna Sofaer and Matt Dibble. Oley, PA: Bullfrog Films.

Spencer, Robert T. 1947. Spanish loanwords in Keresan. *Southwestern Journal of Anthropology* 3 (2): 130–46.

S. S. Papadopulos and Associates Inc. 2000. *Middle Rio Grande water supply study: Prepared for the U.S. Army Corps of Engineers and the New Mexico Interstate Stream Commission.* Bethesda, MD: S. S. Papadopulos and Associates Inc.

Steele, Thomas J. 1994. *Santos and saints: The religious folk art of Hispanic New Mexico.* Santa Fe, NM: Ancient City Press.

———. 2005. *The Alabados of New Mexico.* Albuquerque: University of New Mexico Press.

Sterrett, Robert J., ed. 2007. *Groundwater and wells.* 3rd ed. New Brighton, MN: Johnson Screens.

Stradling, David. 2004. *Conservation in the Progressive era.* Seattle: University of Washington Press.

Stromberg, J. C., V. B. Beauchamp, M. D. Dixon, S. J. Lite, and C. Paradzick. 2007. Importance of low-flow and high-flow characteristics to restoration of riparian vegetation along rivers in arid southwestern United States. *Freshwater Biology* 52:651–79.

Stromberg, J. C., M. K. Chew, P. L. Nagler, and E. P. Glenn. 2009. Changing perceptions of change: The role of scientists in *Tamarix* and river management. *Restoration Ecology* 17 (2): 177–86.

Stromberg, J. C., S. J. Lite, R. Marler, C. Paradzick, P. B. Shafroth, D. Shorrock, J. White, and M. White. 2007. Altered stream flow regimes and invasive plant species: the *Tamarix* case. *Global Ecology and Biogeography* 16:381–93.

Stuart, David E. 2000. *Anasazi America: Seventeen centuries on the road from Center Place.* Albuquerque: University of New Mexico Press.

Stuart, David E., and Rory P. Gauthier. 1986. The riverine period. In *New Mexico in maps*, 2nd ed., ed. Jerry L. Williams, 89–91. Albuquerque: University of New Mexico Press.

Sublette, J. E., M. D. Hatch, and M. Sublette. 1990. *The fishes of New Mexico.* Albuquerque: University of New Mexico Press.

Swetnam, Thomas. 2009. Physical impacts and predictions: What science can tell us. Paper presented at Adaptation to Climate Change in the Desert Southwest: Impacts and Opportunities Conference, Tucson, AZ.

Taylor, William B. 1975. Land and water rights in the viceroyalty of New Spain. *New Mexico Historical Review* 50 (3): 189–212.

Tetra Tech. 2004. *Habitat restoration plan for the middle Rio Grande.* Albuquerque, NM: U.S. Bureau of Reclamation.

Theis, C. V. 1935. The relation between the lowering of the piezometric surface and the rate and duration of the discharge of a well using ground-water storage. *Transactions–American Geophysical Union* 16 (2): 519–24.

———. 1940. The source of water derived from wells: Essential factors controlling the response of an aquifer to development. *Civil Engineering* 10 (5): 277–80.

———. 1941. The effect of a well on the flow of a nearby stream. *Transactions– American Geophysical Union* 22 (3): 734–38.

———. 1953/1991. Outline of ground-water conditions at Albuquerque: Talk given to Chamber of Commerce. In *Short papers on water resources in New Mexico, 1937–1957*, ed. C. V. Theis 19–22. U.S. Geological Survey Open-File Report 91-81. Reston, VA: U.S. Geological Survey.

Thomle, Irwin. 1948. The developmental period in San Luis Valley agriculture: 1898 to 1930. PhD thesis, Northwestern University.

Thomson, B. 2008. Brackish water can't sustain N.M. *Albuquerque Journal*, November 30.

Thorn, C. R., D. P. McAda, and J. M. Kernodle. 1993. *Geohydrologic framework and hydrologic conditions in the Albuquerque Basin, central New Mexico*. Water Resources Investigation Report 93-4149. Albuquerque, NM: U.S. Geological Survey.

Tuan, Yi-fu. 1973. *The climate of New Mexico*. Santa Fe, NM: State Planning Office.

Twitchell, Ralph Emerson, ed. 1914/2008. *The Spanish Archives of New Mexico*. 2 vols. Repr. Santa Fe, NM: Sunstone Press.

U.S. Bureau of the Census. 2009. Resident population and apportionment of the U.S. House of Representatives. http://www.census.gov/dmd/www/resapport/ states/newmexico.pdf (accessed January 29, 2009).

U.S. Bureau of Reclamation. 1951. 1951 annual report, Bureau of Reclamation, Rio Grande Project. In *Bureau of Reclamation, Rio Grande Project reports 1950– 1985*. Albuquerque, NM: U.S. Bureau of Reclamation.

———. 1956. 1956 annual report, Bureau of Reclamation, Rio Grande Project. In *Bureau of Reclamation, Rio Grande Project reports 1950–1985*. Albuquerque, NM: U.S. Bureau of Reclamation.

———. 1972. *Repayment of reclamation projects*. Washington, D.C.: GPO.

———. 1983. *Elephant Butte Reservoir: 1980 sedimentation survey*. Denver, CO: U.S. Bureau of Reclamation.

———. 2005. *Final environmental assessment, San Acacia priority sites, river miles 114 and 113, Albuquerque area office, FONSI, February 28, 2005*. Albuquerque, NM: U.S. Bureau of Reclamation.

———. 2006. *North central Arizona water supply study: Report of findings*. Denver, CO: U.S. Bureau of Reclamation.

U.S. Forest Service. 2003. *Forest insect and disease conditions in the southwestern region, 2003*. Publication R3-03-01. Albuquerque, NM: Forest and Forest Health, U.S. Forest Service.

U.S. Reclamation Service. 1903. *Annual report of the United States Reclamation Service, 1902–1903*, ed. F. H. Newell. H. Doc. 44, 58th Cong., 2nd sess. Washington, D.C.: GPO.

———. 1904a. *Annual report, 1903–1904*. Washington, D.C.: GPO.

———. 1904b. *Personnel of the Service. First conference of engineers of the Reclamation Service*, ed. F. H. Newell. Water Supply and Irrigation Paper 93. Washington, D.C.: GPO.

———.1905. *Second conference of engineers of the Reclamation Service, 1904–1905, at El Paso, TX, and Washington, D.C.*. Washington, D.C.: GPO.

———. 1917. *Project History: Rio Grande Project, New Mexico–Texas.* Unpublished report, U.S. Bureau of Reclamation, El Paso, TX.

Utton, Albert E. 1994. Water in the arid Southwest: An international region under stress. Annual Research Lecture, University of New Mexico.

Van Cleave, M. 1935. Vegetation changes in the middle Rio Grande Conservancy District. MS thesis, University of New Mexico.

Van Denburgh, A. S., and J. H. Feth. 1965. Solute erosion and chloride balance in selected river basins of the western conterminous United States. *Water Resources Research* 1:537–41.

van Mantgem, Philip J., Nathan L. Stephenson, John C. Byrne, Lori D. Daniels, Jerry F. Franklin, Peter Z. Fulé, Mark E. Harmon, et al. 2009. Widespread increase of tree mortality rates in the western United States. *Science* 23:521–24.

Van Riper, C., K. Paxton, and C. O'Brien. 2008. Rethinking avian response to *Tamarix* on the lower Colorado River: A threshold hypothesis. *Restoration Ecology* 16:155–67.

Veatch, William. 2008. Quantifying the effects of forest canopy cover on net snow accumulation at a continental, mid-latitude site, Valles Caldera National Preserve, New Mexico, USA. MS thesis, University of Arizona.

Vierra, Bradley J., ed. 2005. *The Late Archaic across the borderlands: From foraging to farming.* Austin: University of Texas Press.

Villagrá Alcalá, Gaspar Peres de. 1610/1933. *History of New Mexico, 1610.* Trans. Gilberto Espinoza. Comp. F. W. Hodge. Los Angeles: The Quivira Society.

Vivian, R. Gwinn. 1974. Conservation and diversion: Water-control systems in the Anasazi Southwest. In *Irrigation's impact on society*, ed. T. Downing and M. Gibson, 95–112. University of Arizona Anthropological Papers 25. Tucson: University of Arizona Press.

———. 2004. Puebloan farmers of the Chacoan world. In *In search of Chaco: New approaches to an archaeological enigma*, 7–13. Santa Fe, NM: School of American Research Press.

Vlasich, James A. 1980. Transitions in Pueblo agriculture, 1938–1948. *New Mexico Historical Review* 55:25–46.

———. 2005. *Pueblo Indian agriculture.* Albuquerque: University of New Mexico Press.

Walker, Arthur L., and Albert S. Curry. 1928. The status of land and capital in the Elephant Butte Irrigation District. *Journal of Land and Public Utility Economics* 4:75–84.

Ward, Frank A., and Manuel Pulido-Velazquez. 2008. Water conservation in irrigation can increase water use. *Proceedings of the National Academy of Sciences* 105 (47): 18215–20.

Warren, Louis S. 1997. *The hunter's game: Poachers and conservationists in twentieth century America*. New York: Yale University Press.

Warren, Nancy Hunter. 1987. *Villages of Hispanic New Mexico*. Santa Fe, NM: School of American Research Press.

Weber, Matthew A., and Steven Stewart. 2009. Public values for river restoration options on the middle Rio Grande. *Restoration Ecology* 17 (6): 762–71.

Welsh, Michael E. 1987. *U.S. Army Corps of Engineers, Albuquerque District, 1935–1985*. Albuquerque: University of New Mexico Press.

Westphall, Victor. 1983. *Mercedes Reales: Hispanic land grants of the upper Rio Grande region*. Albuquerque: University of New Mexico Press.

Whiteley, Peter Michael. 1982. Third Mesa Hopi social-structural dynamics and socio-cultural change: The view from Bacavi. PhD thesis, University of New Mexico.

Wigley, T. M. L., and P. D. Jones. 1985. Influences of precipitation changes and direct CO_2 effects on streamflow. *Nature* 314:149–52.

Wilber, Charles Dana. 1881. *The great valleys and prairies of Nebraska and the Northwest*. Omaha, NE: Daily Republican Printing.

Wilcox, Bradford P., and M. Karl Wood. 1989. Factors influencing interrill erosion from semiarid slopes in New Mexico. *Journal of Range Management* 42:66–70.

Wilcox, Michael. 2010. Marketing conquest and the vanishing Indian: An indigenous response to Jared Diamond's archaeology of the American Southwest. In *Questioning collapse: Human resilience, ecological vulnerability, and the aftermath of empire*, ed. Patricia A. McAnany and Norman Yoffee, 113–41. New York: Cambridge University Press.

Wills, W. H. 1992. Plant cultivation and the evolution of risk-prone economies in the prehistoric American Southwest. In *Transitions to agriculture in prehistory*, ed. A. G. Gebauer and T. D. Price, 153–76. Madison, WI: Prehistory Press.

Workman, E. J., and S. E. Reynolds. 1950. Electrical phenomena occurring during the freezing of dilute aqueous solutions and their possible relationship to thunderstorm electricity. *Physics Review* 78 (3): 254–59.

Wozniak, Frank E. 1997. *Irrigation in the Rio Grande valley, New Mexico: A study and annotated bibliography of the development of irrigation systems*. General Technical Report RMRS-P-2. Fort Collins, CO: Rocky Mountain Research Station, U.S. Forest Service.

Yablonski, Brian. 2004. Valles Caldera National Preserve: A new paradigm for federal lands? *PERC Reports* (December): 3–5.

Young, Michael. 2009. Looking to the future in view of past experience: Economic, policy and legal perspectives. Panel presentation at Adaptation to Climate Change in the Desert Southwest: Impacts and Opportunities, Tucson, AZ, January 22–23.

Index

Page numbers in italic text indicate illustrations.

Haase, Steve, 176–77
Hall, B. M., 92–93
Hansen, Steve, 155–58
Hantush, Mahdi, 133, 213n18
Hawley, John, 176–77
Hay, John, 91
Hayden, Ferdinand, 55
headgates, 28, 29, 30, 45, 45–46
Hernandez, B. C., 119
Hill, Raymond A., 125–26. *See also* Rio
  Grande Compact
Hispanics, 4, 24, 212n40; irrigation
  communities of, 38–52, 160; MRGCD
  opposed by, 113–19, 122; Pinchot and
  Newell policies impacting, 95–96;
  Pueblo culture compared to, 49–51;
  stream adjudication involving, 165;
  subsistence farmers, 67–79. See also
  acequia
Hohokam culture, 26–29, 172
Homol'ovi IV (Arizona), 206n20
Hopi, 27; migrations of, 206n20; water's
  significance to, 34–36, 35
Hosta Butte, 60–61
hunting, Paleoindian big game, 24–25,
  206n6
hydrologic science, 81–85, 82, 83; global
  hydrological system, 19–20; Rio
  Grande hydrological system, 20–21
hydroponic forage greenhouses, 184

importation, of water, 144, 152–53, 166–
  68, 175, 180–82
instream flow, 150–51, 156–57
International Boundary and Water
  Commission, 84–85
Interstate Stream Commission, City of
  Albuquerque, 154
inverted pyramid, pricing structures,
  189–91
irrigation: *acequia*, 37–52, 42, 45, 48, 83,
  84, 86, 102; aesthetic and ecological
  benefits of, 184; ancient Southwestern,
  28–29, 207n23, 207n45; Boyd's Rio
  Grande Dam and Irrigation Company,

88–92, 89; canal, 30, 67–70, 79; Cochiti
  Diversion, 1, 3, 6, 6–10, 9, 142, 146,
  152–53, 205n15; at Cochiti Pueblo,
  3; Elephant Butte Dam study of,
  183; Hispanic community coalesced
  around, 38–52, 160; Iberian, 4;
  National Irrigation Congress, 92–93;
  Navajo Irrigation Project, 141; Rio
  Grande status as irrigation ditch, 86;
  runoff, 27, 30, 43, 64–65, 111, 169–70,
  173, 174
Isabella (queen of Spain), 164
Islamic irrigation institutions, 29
Isleta Pueblo, 147, 166
Israel, 177

Jemez Canyon Dam: as marvel or horror,
  152–53; siltation problem at, 141–42
Jemez Mountains, 2, 18
Jemez River, 33
jetty jacks, 144; removal of, 168, 200
July Feast Days, 8
juniper, 60–61

katsina (kachina) ceremonies, 34–36
Kearny, Stephen Watts, 53–54
Keresan language, 4
Kuwaninvaya, Cedric, 206n20

La Bajada, 2, 3, 39
Ladd, Edmund, 31
Laguna Pueblo, 31
La Joya (New Mexico), 48
La Mesa (New Mexico), 72
land grants: Acequia del Molino, 40–43,
  42, 190; forest resources locked up in,
  61; Gallegos land grant, 112–13, 117–18,
  120–21; overlapping, 5; Pueblo obtains
  Spanish, 5; Santa Barbara Tie and Pole
  Company obtaining, 64–65; U.S. legal
  maneuvering of, 62–64
La Plata Mountains, 60–61
La Plata River, 123
Las Cruces (New Mexico), 136
Las Vegas (Nevada), 167–68

Thomas, Jack, 176
Thomson, Bruce, 177–78
Traveler's Insurance Company, 69
Treaty of Guadalupe Hidalgo, 61–62;
   stream system adjudication and,
   164–66
trout, 149
Tucson (Arizona), 167–68, 190
Tuggle, Benjamin, 155
Turner, Bill, 189

Udall, Tom, 150–51
United Nations Intergovernmental Panel
   on Climate Change, 172
United States: Cochiti Pueblo and, 5–9;
   global warming impact on, 171–74,
   201–2; legal maneuvering of, 62–64;
   New Mexico annexation by, 5, 53–55;
   New Mexico authority in Washington,
   D.C., 59–60; Treaty of Guadalupe
   Hidalgo by, 61–62, 164–66. *See also
   specific agency, city, or state*
University of Adelaide (Australia), 192
University of New Mexico, 147–50,
   164–66, 178, 202; Water Resources
   Program, 156
Upper Rio Grande Water Operations
   Model (URGWOM), 169–70
URGWOM. *See* Upper Rio Grande
   Water Operations Model
USGS. *See* Geological Survey, U.S.
Utes, 68
Utton, Al, 202, 218n6

Vale of Siddim, 76–77
Valles Caldera, 159–60, 192–94
valuation, water, 189–91, 193–94
vegetation, Rio Grande, 20–21
vinculaciónes (obligations of serf to
   king), 50–51

volcanoes, 11, 13, 15, 18

waffle gardens, 27, 28
wastewater, 78–79, 147, 175
water buffaloes (long-time state
   administrators), 146
waterlogged farmland, 112–13, 117, 121,
   202
water markets, 184–88
water supply: attitude of reciprocity
   toward, 200–201; conservation of, 182–
   83; desalination to augment, 176–79;
   global warming impacting, 171–74,
   201–2; importing of, 144, 152–53, 166–
   68, 175, 180–82; management of, 45–47,
   84, 191–95; population limit based on,
   181, 195–96; reallocation of, 183–84;
   saline aquifers to augment, 176–77
wells, 28, 138, 174; groundwater pumped
   from, 137; Hantush and Theis on
   hydraulics, 133–34, 213n18; Sierra
   Waterworks Company proposal for,
   180–81; Theis on high capacity, 134–35
wetlands, 78, 168
wheat crop, 108
White Rock Canyon, 35
Wild Earth Guardians, 153
willow trees, 148, 163. See also *bosque*
Wootton, Richens ("Uncle Dick"), 57
Workman, E. J., 213n18
Wyatt the Wolf, 153
Wyoming, 101

Yale University, 95
Young, Michael, 192, 194–95

zansas (irrigation duties), 29
Zuni Pueblo, 27–28, 28, 34–35, 207n45
Zuni River, 33

REINING IN THE RIO GRANDE
Design and composition: Karen Mazur
Set in Minion Pro with Scala Sans display
Printed by Thomson Shore, Inc., Dexter, Michigan